OSHA's Respiratory Protection Standard

A Proven Written Program for Compliance

Mark McGuire Moran

Government Institutes, Inc.
Rockville, Maryland

Government Institutes, Inc.
4 Research Place, Suite 200
Rockville, Maryland 20850

Copyright © 1996 by Government Institutes. All rights reserved.

99 98 97 96   5 4 3 2 1

No part of this work may be reproduced or transmitted in any form or by any means, electronic or mechanical, including photocopying, recording, or any information storage and retrieval system, without permission in writing from the publisher. All requests for permission to reproduce material from this work should be directed to Government Institutes, Inc., 4 Research Place, Suite 200, Rockville, Maryland 20850.

The author and publisher make no representation or warranty, express or implied, as to the completeness, correctness or utility of the information in this publication. In addition, the author and publisher assume no liability of any kind whatsoever resulting from the use of or reliance upon the contents of this book.

ISBN: 0-86587-501-4

Printed in the United States of America.

TABLE OF CONTENTS

INSTRUCTIONS vi

RESPIRATORY PROTECTION PROGRAM 1

I. INTRODUCTION 2
II. ADMINISTRATION 2
III. USE OF RESPIRATORS 3
    Respirator User Card 6
    Medical Questionnaire For Respirator Program 8
IV. SELECTION OF RESPIRATORS 9
V. EMPLOYEE TRAINING, INSTRUCTION AND DISCIPLINE 13
VI. INSPECTION, CLEANING, STORAGE & REPAIR 15
VII. EMERGENCY-USE RESPIRATORS 16
    Gas Mask Table 18
VIII. PROGRAM EVALUATION 19
    Respirator Training Record 20

APPENDIX A: 29 CFR 1910.134—RESPIRATORY PROTECTION 21

**APPENDIX B: OSHA INSTRUCTION CPL 2-2.54—
RESPIRATORY PROTECTION PROGRAM MANUAL** 29

    Chapter I—RESPONSIBILITIES 34
    Chapter II—RESPIRATORY PROTECTION PROGRAM ADMINISTRATION . 39
    Chapter III—RESPIRATOR SELECTION 43
    Chapter IV—TRAINING 51
    Chapter V—FIT TESTING 52
    Chapter VI—MAINTENANCE 65
    Appendix A—SAMPLE CHECKLIST AND INSPECTION FORM
        FOR MSA SELF-CONTAINED BREATHING APPARATUS 67
    Appendix B—SAMPLE INSPECTION AND MAINTENANCE PROCEDURES
        FOR SURVIVAIR SELF-CONTAINED BREATHING APPARATUS 72

INDEX 77

INSTRUCTIONS FOR USE OF
THE RESPIRATORY PROTECTION PROGRAM

1. The accompanying program covers everything that is required by the written program requirements of the OSHA respiratory protection standard, 29 C.F.R. §1910.134. You only need to insert your company name on the cover page and, on page 2, the name of the person in your company chosen to oversee the program. Before doing so, however, observe the instructions that follow.

2. We have added a copy of *OSHA's Respiratory Protection Program Manual* as part of this written Program. This is a compliance document that OSHA Inspectors use to check written respiratory protection programs. It is recommended that the reader look through this program to see if their company meets the requirements that are listed in this manual.

3. Read every word of the Program. It explains what the standard requires. It also explains various types of respirators, how they work, what they do and when they should be worn. Some of the Program's provisions can be eliminated, changed or modified in order to reflect your actual work practices. A copy of the OSHA standard is attached as Appendix A. If you choose to make changes in the written program, you should check the standard to make sure you are not thereby violating an OSHA requirement.

4. The Program contains two forms: A respirator user card and a medical questionnaire. Neither of them is required by the standard so you can eliminate them from the Program if you choose. You should be aware, however, of §1910.134(b)(9) of the standard. It states that: "Persons should not be assigned to tasks requiring use of respirators unless it has been determined that they are physically able to perform the work and use the equipment." Because it uses the word "should," it is not mandatory. Observing this provision is good practice and the form provided on can be helpful.

5. There is another form included at the end of the Program for the purpose of documenting the respirator training that you provide for employees. Although the standard does not require the documentation of training, keeping such a record can be helpful in many ways.

6. There are a number of additional OSHA standards that regulate specific substances and also include respirator provisions. Consequently, if your employees are exposed to those substances, you should check the applicable respirator requirements and adapt your Written Respiratory Protection Program accordingly. See, for example, 29 C.F.R. §§1910.1001(g)(3); 1910.1101(d)(2)(iv) and 1926.58(h)(3) [*asbestos*]; 1910.1017(g)(3) [*vinyl chloride*]; 1910.1018(h)(4) [*arsenic*]; 1910.1025(f)(4) [*lead*]; 1910.1028(g)(3) [*benzene*]; 1910.1029(g)(3) [*coke oven emissions*]; 1910.1043(f)(3) [*cotton dust*]; 1910.1044(h)(3) [*1,2, dibromo-3-chloropropane*]; 1910.1045(h)(3) [*acrylonitrile*]; 1910.1047(g)(3) [*ethylene oxide*] and 1910.1048(g)(3) [*formaldehyde*].

7. There are also a number of material safety data sheets (MSDSs) that include respirator provisions. They are part of the precautions recommended by the manufacturer of the product listed on the MSDS (such as paint, for example). Some such MSDSs state that a respirator is required "when using this product." Others, specify a particular type of respirator. Still other MSDSs require a respiratory function test. Numerous varieties of precautions like these appear on various MSDSs. Although complying with MSDS precautions is not an OSHA requirement, failure to observe these precautions could create difficulties for you if they are not observed *and* an employee is overcome or contracts an illness. Consequently, observance of MSDS respirator recommendations has been made a part of the accompanying Program. See paragraph 1 of the "Use of Respirators" section.

8. Respirator manufacturers and suppliers also provide instructions on the proper use and care of their particular brand of respirator. These instructions should be observed and a rule to that effect has been written into the accompanying Program. See paragraph 6 of the section on Employee Training, Instruction and Discipline.

9. If it is possible to do so—and thought appropriate—reproduce your written respiratory protection program and distribute it to each affected employee. You will notice that it is written in a style that is addressed to each affected employee, gives advice to them and contains work rules for them. There could be occasions when those rules might be useful to the company—provided employees are aware of them. Consequently, if you do not choose to provide employees with copies of your Program, then the rules for their conduct that are included in the Program should be brought to the employees attention through some other methods, such as training programs or safety meetings.

10. Section VII of the Program (pp. 16-19) is confined to emergency-use respirators. If you do not have any such respirators, delete that entire section.

11. **—WARNING—**

You must observe the rules and procedures included in your written Respiratory Protection Program, as well as, all other written documents or instruments adopted to implement job safety and health provisions. Failure to observe these rules and procedures can result in severe consequences. For example, willful OSHA violations are frequently based upon an allegation that the *employer* had written work rules but that his or her work *practices* were *different*—thereby indicating an intent to deceive. Consequently, if the procedures or rules you choose to follow are *different* from those included in this Respiratory Protection Program, you must make appropriate changes in that Program before you add your name to it.

ADDITIONAL INFORMATION

Moran Associates stands behind this Respiratory Protection Program. If, at any time, you have a question, *want* additional guidance, or wish to offer a comment or suggestion, please contact *Moran Associates* by telephone at (904) 278-5155.

> *NOTE: Purchasers of this Respiratory Protection Program will not be charged any fee by Moran Associates for answering questions relating to this Program or providing assistance in putting it into effect.*

RESPIRATORY PROTECTION PROGRAM OF

_____ COMPANY NAME

I. INTRODUCTION

1. This program is designed to provide information and instruction upon the proper use, care, inspection, cleaning, repair and storage of respirators. It is to be read and implemented in conjunction with applicable OSHA standards including 29 C.F.R. §1910.134 (respiratory protection), and any other OSHA standard regulating the use of respirators that may apply to our work conditions and practices; a copy of §1910.134 is attached. Although the OSHA standard is part of the "general industry" standards, it has been designated by OSHA as applicable to construction work and therefore applies to both general industry and construction.

2. We have established a respiratory protection program in order to coordinate the use and maintenance of respiratory protective equipment and to (a) reduce employee exposure to toxic chemical agents, and (b) allow employees to work safely in hazardous work environments.

3. Each employee who has been issued a respirator and each respirator-wearer, must be aware of the need to observe the provisions of this program.

4. Whenever any employee becomes aware of a problem or potential trouble of any kind related to respirator use, he or she must notify his or her supervisor at once.

5. If, at any time, there is something in this program, or anything about the respirator or its use, that is not understood, or anything an employee wants to have further explained, he or she must immediately notify his or her supervisor. That supervisor must then see to it that the employee is provided with the necessary instruction, training, information or protection.

II. ADMINISTRATION

1. _____ has been designated as the company person responsible for this respiratory protection program.

2. They have authority to make decisions and implement changes in our respirator program when necessary.

3. They will have the following responsibilities:
    (a) Supervision of respirator selection procedure;

(b) Establishment of respiratory equipment training programs for employees;

(c) Establishment of a continuing program of cleaning and inspection of respiratory equipment;

4. (a) Designation of proper storage areas for respiratory equipment;

(b) Establishment of issuance and accounting procedures for uses of respiratory equipment;

(c) Establishment of medical screening program/procedures for employees assigned to wear respiratory equipment;

(d) Establishment of a periodic inspection schedule of those workplace/conditions that require respiratory equipment in order to determine exposure and/or changing situations;

(e) A continuing evaluation of the above aspects to assure their continued functioning and effectiveness.

(f) Regular inspection and evaluation of this Respiratory Protection Program in order to determine its continued effectiveness.

(g) Keeping advised of all applicable OSHA respiratory protection requirements and advising affected supervisors and employees thereof as necessary.

5. Any employee who has questions or problems with respirators or their use must immediately notify his immediate supervisor. If such supervisor cannot resolve the question or problem, they will refer the matters to the person in charge of our respirator program (see paragraph No. 1, above).

III. USE OF RESPIRATORS

1. It is mandatory that employees wear the appropriate respirator when working where the air contains regulated substances in concentrations exceeding the permissible exposure limit (PEL), and whenever there is potential exposure to a contaminant substance for which its material safety data sheet (MSDS), prescribes respirator use. The use of respirators is also required when working at any place where there exists any one of the hazards listed in the table at the end of Part IV (Selection of Respirators) of this Program.

2. All company supervisors will maintain continued surveillance of work conditions in all places where employees for whom they are directly responsible work, as well as employee exposures and stress—in order to determine if any additions to, or changes in, respirator use requirements are needed.

3. The supervisor will promptly notify employees of changes whenever they are needed.

4. This respirator program, the instructions accompanying the respirator, the applicable OSHA regulations and the precautions stated in the MSDS for each of the substances being protected against must be observed by each user of a respirator.

5. No employee will perform a job that requires respirator use, or be present at any place where respirators are necessary unless all provisions of this Respirator Program are observed.

6. Any employee performing such a job, or present in such a place, who is wearing a respirator must immediately cease his work, leave the area and report the matter to his supervisor whenever any one of these conditions exists: (a) dizziness, difficulty in breathing, or other physical stress or disorder; (b) damage to, or ineffectiveness of, the respirator being worn; (c) the smell or taste of any contaminant, or any unfamiliar smell or taste or other such sensation that troubles or concerns any employee, or (d) lack of the respirator training and instruction required under this Program, or the absence of any other requirement of this Program.

7. Surveillance of conditions in the work area and degree of worker exposure or stress (combination of work rate, environmental conditions, and physiological burdens of wearing a respirator) must be maintained at all times.

8. Changes in operating procedures, temperature, movement of air, humidity and work practices may influence the concentration of a substance in the work area atmosphere. These factors may necessitate periodic monitoring of the air contaminant concentration. If testing is undertaken, it should continue in order to assure that the contaminant exposure has not risen above the maximum protective capability of the respirators being used.

9. Employees using self-contained breathing apparatus or supplied-air respirators in *confined spaces*, where the air is or may be "immediately

dangerous to life or health," must wear safety harnesses and lifelines. A second person equipped with complete protective gear must be standing by ready to help if the first worker gets into trouble. Communications (visual, voice, or signal line) must be maintained with all persons present. Precautions must be taken so that in the event of an incident, at least one person will be unaffected and have the proper rescue equipment to be able to assists the others in an emergency situation.

10. Only those individuals who are medically able to wear respiratory protective equipment will be issued one. No employee will wear a respirator unless he or she is medically able to do so.

11. No one will be permitted to use a respirator unless they are physically able to perform the work while wearing the respirator.

12. Prior to any respirator use, each employee must fill out a medical questionnaire (a copy of which is attached), and be examined by appropriate medical personnel for vital capacity, circulatory problems and fitness to wear the respirator.

13. This fitness requirement will be reviewed once a year but if any respirator user's physical ability changes at any time, they must notify their supervisors at once and cease respirator use, until medical approval for resuming respirator use has been obtained.

14. The examining physician will determine what health and physical conditions are pertinent to an employee's respirator usage.

15. Whenever there is evidence of employee exposure to toxic substances while wearing a respirator, it will be followed-up with appropriate surveillance of work area conditions to determine if there is any relationship to inadequate respiratory protection, or a need for additional or other kinds of controls.

16. All individuals who are assigned to wear respiratory protective equipment will be provided respiratory protective equipment for their exclusive use.

17. A system of respiratory wearer cards and journals has been established to facilitate the accounting of users and equipment. The following user card and journal scheme has been adopted.

RESPIRATOR USER CARD

CARD NUMBER: _____

NAME: _____

OPERATION: _____

CONTAMINANTS/HAZARD PROCESS: _____
_____

RESPIRATOR TYPE: _____

DATE OF ISSUANCE: _____

DATE OF EXPIRATION: _____

APPROVED BY: _____

18. Each respirator user must receive fitting instructions including demonstrations and practice on how the respirator should be worn, adjusted, and how to determine if it fits properly.

19. Although respirators are designed for maximum efficiency, they cannot protect the wearer without a tight seal between the facepiece and wearer. Beards and other facial hair can substantially reduce the effectiveness of a respirator. The absence of dentures can seriously affect the fit of a facepiece. To assure proper protection for a facepiece, it must be checked by the wearer each time he or she puts on the respirator.

20. Corrective glasses worn by employees also present a problem when fitting respirators. Special mountings to hold corrective lenses inside full facepieces are available. If corrective lenses are needed, the facepiece and lenses must be fitted by a qualified individual to provide good vision, comfort and proper sealing.

21. Contact lenses should not be worn while wearing a respirator in a contaminated area. Foreign bodies or contaminants that penetrate the respirator may get into the eyes and cause severe discomfort causing the wearer to remove the respirator.

22. Full facepieces, half masks, and quarter masks have different fitting characteristics. Of the several brands of each style marketed, each has a different size and fitting characteristic; no respirator will fit everyone.

23. Any employee who finds that they cannot obtain a proper fit with his or her respirator, must notify their supervisors immediately.

24. Upon learning of any respirator's improper fit, the supervisor will not permit the employee to work in any area where respirator's are required, *until* the employee is equipped with a proper-fitting respirator.

25. The effectiveness of the facepiece fit of a respirator can be tested two ways—qualitatively and quantitatively:

    (a) Qualitative fit testing involves the introduction of a harmless, odorous or irritating substance into the breathing zone of the wearer. Not detecting the substance indicates a proper fit.

    (b) Quantitative fit testing offers the most accurate, detailed information on respirator fit. It involves the introduction of a harmless aerosol to the wearer while he or she is in a test chamber. While the wearer performs exercises that could induce facepiece leakage, the air inside and outside the facepiece is then measured for the presence of the harmless aerosol.

26. The supervisor of each respirator wearer is responsible for ensuring that the appropriate facepiece fit test has been conducted and that the result of such test has indicated a proper fit.

27. All employees who will or may be required to wear respirators must complete the following questionnaire:

MEDICAL QUESTIONNAIRE FOR RESPIRATOR PROGRAM

Please indicate if you have any of the following: (if you answer any items "yes," please explain below).

	YES	NO
Emphysema	___	___
Chronic obstructive pulmonary disease	___	___
Bronchial Asthma	___	___
Pneumoconiosis	___	___
Coronary artery disease	___	___
Cerebral blood vessel disease	___	___
Severe or progressive hypertension	___	___
Epilepsy, grand mal or petit mal	___	___
Anemia	___	___
Diabetes	___	___
Punctured eardrum	___	___

	YES	NO
Breathing difficulty when wearing a respirator	_____	_____
Claustrophobia or anxiety when wearing a respirator	_____	_____

Describe the nature of illnesses or any medications: _____

_____

_____

_____

_____

Your Name: _____

Date: _____
(Signature)

IV. SELECTION OF RESPIRATORS

1. Choosing the right respiratory protection equipment involves several steps:

    (a) Determining what the hazard is and its extent;
    (b) Choosing equipment that is certified for the function; and
    (c) Assuring that the device is performing the function it is intended to do.

2. Proper selection of respirators must be made according to the OSHA requirements set forth in 29 C.F.R. §1910.134(c) and the American National Standards Institute publication *Practices for Respiratory Protection, ANSI Z88.2-1969*. All respiratory protective devices must be approved for the contaminant or situation to which the employee is exposed by either the Mine Safety and Health Administration, U.S. Department of Labor, or the National Institute for Occupational Safety and Health, U.S. Department of Health and Human Services.

3. In addition to the foregoing, there are substance-specific OSHA standards that require additional criteria for respirator selection (for example, 29 C.F.R. §1910.1025(f)(2) [airborne lead]). All such requirements of each applicable OSHA standard must be observed.

4. Chemical and physical properties of the contaminant, as well as the toxicity and concentration of the hazardous material and the amount of oxygen present, must be considered in selecting the proper respirators.

5. The nature and extent of the hazard, the work rate, the area to be covered, mobility, work requirements and conditions, as well as the limitations and characteristics of the available respirators, also are selection factors that must be considered.

6. Although there are many kinds of breathing equipment that are used for protection, there are two basic types—air purifying and atmosphere-supplying respirators:

    (a) *Air-Purifying Respirators* are designed to remove harmful substances from the air. They range from simple disposable masks to sophisticated positive-pressure, blower-operated respirators. Air-purifying respirators may not be used in an oxygen deficient atmosphere or under immediately-dangerous-to-life-or-health conditions.

    (b) *Atmosphere-Supplying Respirators* are designed to provide air from a clean source outside of the contaminated work area. They range from air-line respirators, self-contained breathing apparatus (SCBA) to complete air-supplied suits.

7. The time needed to perform a given task usually determines the length of time for which respiratory protection is needed, including the time necessary to enter and leave a contaminated area.

8. A self-contained breathing apparatus, gas mask, or chemical-cartridge respirator provides respiratory protection for relatively short periods, while the air-line respirator provides protection for as long as the facepiece is supplied with adequate respirable air.

9. Particulate-filter respirators can provide protection for long periods without need of filter replacement *only* if the total atmospheric particulate concentration is low.

10. For protracted periods of use, air-line respirators offer the advantage of longer use in high dust loading areas and avoid the need to be concerned about the sensory warning properties of the airborne toxic materials. Those respirators also cause less discomfort than air-purifying respirators because the wearer need not overcome filter resistance in order to inhale.

11. Some type of warning on remaining service life is available for all self-contained breathing apparatus and on some gas masks. It may be a pressure gauge or timer with an audible alarm for self-contained breathing apparatus or a window-indicator in the canister. The respirator user must understand the operation and limitations of each type of warning device. Most other gas masks and chemical-cartridge respirators have no indicator of remaining service life. Therefore, it is important that canisters and cartridges be changed according to the manufacturer's directions.

12. Air-purifying respirators present minimal interference with the wearer's movement. Supplied-air respirators with trailing hoses severely restrict the area the wearer can cover and present a potential hazard where the trailing hose can come in contact with machinery. A self-contained breathing apparatus—a unit that includes a back-mounted compressed-air cylinder—presents a size and weight penalty that may restrict climbing and movement in tight places.

13. The wearer's work rate determines the volume of air breathed per minute, maximum inspiratory flow rate, and the tolerable inhalation and exhalation breathing resistance. The respiratory minute volume is of great significance in self-contained and air-line respirators operated from cylinders, since it determines their operating life. Useful life under moderate working conditions may be significantly less than that under rest conditions.

14. Peak airflow rate is important in the use of constant-flow air-line equipment. The air-supply rate should always be greater than the peak inspiratory flow rate to maintain the respiratory enclosure under positive pressure.

15. High breathing resistance of air-purifying respirators under conditions of heavy work can result in distressed breathing.

16. A person working in an area of high temperature is under stress. Additional stress resulting from use of a respirator should be minimized by using a respirator with minimum weight and breathing resistance.

17. Unless there exists a specific OSHA standard containing different requirements that are applicable, respirators will be selected according to the matters stated above and the following table.

	Hazard	Respirator
(a)	Oxygen Deficiency	
—	Immediately dangerous to life*	Any self contained breathing apparatus.
—	Not immediately dangerous to life or health	Self-contained breathing apparatus. Any supplied-air respirator.
(b)	Gas and Vapor Contaminants	
—	Immediately dangerous to life or health*	Positive-pressure, self-contained breathing apparatus. Combination positive-pressure supplied air-respirator and self-contained air supply.
—	Not immediately dangerous to life or health	Any supplied air-respirator. Gas mask. Chemical cartridge respirator.
(c)	Particulate Contaminants	Any supplied-air respirator including abrasive blasting respirator. Powered air-purifying respirator equipped with efficiency filters. Any air-purifying respirator with particulate filter.
(d)	Gaseous and Particulate Contaminants	
—	Immediately dangerous to life health*	Positive-pressure, self-contained breathing apparatus. Combination positive-pressure supplied-air respirator and auxiliary self-contained air supply.

	Hazard	Respirator
—	Not immediately dangerous to life or health	Any supplied-air respirator. Gas mask. Chemical-cartridge respirator.
(e)	Escape from contaminated atmosphere which may be immediately dangerous to life or health	Any self-contained breathing apparatus. Gas mask. Combination air-line respirator with escape self-contained breathing apparatus.
(f)	Firefighting	Any positive-pressure, self-contained breathing apparatus.

*Note: "Immediately dangerous to life or health" is any condition that poses either an immediate threat to life or health or an immediate threat of severe exposure to contaminants, such as radioactive materials, that are likely to have adverse delayed effects on human health.

V. EMPLOYEE TRAINING, INSTRUCTION AND DISCIPLINE

1. Every employee who is required to wear a respirator must know how to wear it, care for it, adjust it and determine if it fits properly and provides the appropriate protection.

2. Supervisors will provide their employees with needed respirator training and instruction.

3. Such instruction and training will be given to any employee under the supervisor's direct and immediate control if the employee has not already received it, or if their prior training or instruction did not satisfy OSHA requirements, or if any doubts or questions exist about respirator use or any of the matters mentioned in this Program.

4. Additional training—on a daily basis if necessary—will be provided by each supervisor whenever it is needed to protect the health and safety of employees.

5. Each respirator wearer will be given an opportunity (if he has not already had one), to handle the respirator, have it fitted properly, test its facepiece-to-face seal, wear it in normal air for a period long enough to gain familiarity with it, have it fit-tested as required by the applicable OSHA regulation and to wear it in a test atmosphere.

6. Each respirator is accompanied by its own set of instructions for proper use, care and protection as well as its limitations. They are printed in (or on) the respirator box, bag or container. Those instructions must be observed.

7. Each respirator user must read and abide by those instructions.

8. Any employee who does not understand the respirator instructions, or cannot observe them, must immediately ask their supervisors for assistance.

9. Any employee who has not been provided with all of the training and instruction set forth above, or who at any time is unsure about respirator use, care or protection, or has any problems or difficulties with work while wearing a respirator, must tell that to their supervisors at once so that they can be provided with the proper training and instruction.

10. Failure to follow all instructions and training on respirator use, care and protection, and/or failure to wear the respirator during all times of exposure, can reduce respirator effectiveness and result in sickness or death. The vapors and mists that can be dangerous to health include some so small no one can see or detect them.

11. It is vital to each employee's health that the respirator training and instruction be observed—and it is vital to each employee's job.

12. Appropriate discipline will be given to any employee who fails to observe any part of our respirator program.

13. Persons who provide respirator training and instruction must make sure that a written record is made as provided below.

14. The form entitled "Respirator Training Record" (attached to this Program) will be executed whenever respirator training is conducted.

VI. INSPECTION, CLEANING, STORAGE & REPAIR

1. Each employee who has finished wearing a disposable respirator or a respirator that is to be used only once, will place the respirator in the appropriate trash or disposal container. It will not be taken from the premises for additional use or used a second time under any circumstances.

2. Those respirators that are routinely used will be regularly cleaned and disinfected by the respirator user. That must be done as frequently as necessary to ensure that it provides proper protection to its wearer.

3. A cleaning and disinfecting solution is provided for use in the cleaning process. It is located adjacent to each respirator storage facility.

4. No one should ever use a respirator that has previously been used by another person.

5. Before putting a respirator on, the user will inspect it for defects and cleanliness. That must be done each and every time a respirator is put on.

6. The respirator must be inspected again after taking it off before putting it in storage.

7. Each respirator that is not routinely used, but is kept ready for emergency use, will be inspected after each use and at least monthly in order to assure that it is in satisfactory working condition.

8. An employee must never wear an unclean respirator or a respirator that is defective in any way.

9. Employees must report any instance of defective or ineffective respirators to their supervisors immediately, including the use of an unclean respirator.

10. Respirator inspection will include a check of the tightness of connections and the condition of the facepiece, headbands, valves, connecting tube, and canisters. Rubber or elastomer parts will be inspected for pliability and signs of deterioration. Stretching and manipulating rubber or elastomer parts with a massaging action will keep them pliable and flexible, and prevent them from taking a set during storage.

11. Employees who do not know how to properly inspect their respirator, must ask their supervisors for assistance.

12. The company will regularly collect and inspect all respirators for defects (including a leak check), to make sure that they are performing up to standard and are providing the proper protection to the wearer.

13. Respirator repairs or part replacement will only be done by experienced persons with parts designed and approved for that particular respirator.

14. No attempt will be made to replace components or to make adjustment or repairs beyond the manufacturer's recommendations.

15. Reducing or admission valves or regulators will be returned to the manufacturer or to a trained technician for adjustment or repair.

16. When not in use, each respirator will be stored to protect against dust, sunlight, heat, extreme cold, excessive moisture, or damaging chemicals.

17. Clearly marked storage places have been assigned to each person who has been issued a respirator. Each person must store their respirator in its proper place and in the correct manner.

18. Respirators must be stored so that the facepiece and exhalation valve rest in a normal position, and that the function will not be impaired by the elastomer setting in an abnormal position.

19. Dust respirators must be placed in clean plastic bags.

20. Respirators will never be stored in such places as lockers or tool boxes unless they are in clean carrying cases or cartons, and the cleaning and storage conditions listed above can be assured.

21. No employees will ever remove a respirator from the premises unless directed to do so by their immediate supervisors.

VII. EMERGENCY-USE RESPIRATORS

1. The provisions of this section of the Program are confined to respirators that are kept for emergency use only, such as gas masks and self-contained breathing apparatus.

2. The foregoing instructions for proper use, care, inspection, and storage of respirators will also apply to emergency-use respirators unless inconsistent with the provisions included in this section of the Program.

3. The instructions for proper use, care, inspection and storage of each emergency-use respirator must be observed at all times. Those instructions accompany each such respirator. They are frequently mounted inside the carrying case lid.

4. A properly worded label is the primary means of identifying a gas mask canister. The secondary means of identifying a gas mask canister is by color code.

5. Each person who either issues or uses any gas mask falling within the scope of this section, will make sure that each gas mask canister purchased or used is properly labeled and colored in accordance with the requirements set forth in this section of our Program, and that the labels and colors are properly maintained at all times thereafter, until the canisters have completely served their purpose.

6. The following must appear in bold letters on each canister:

   Canister For: _____
   (Name for atmospheric contaminant)
   or
   Type N Gas Mask Canister

7. In addition, essentially the following wording must appear beneath the appropriate phrase on the canister label: *"For Respiratory Protection in atmospheres containing not more than* _____ *percent by volume of* _____*."*

8. Canisters having a special high-efficiency filter for protection against radionuclides and other highly toxic particulates must be labeled with a statement of the type and degree of protection afforded by the filter. The label will be affixed to the neck end of, or to the gray stripe which is around and near the top of the canister. The degree of protection will be marked as the percent of penetration of the canister by a 0.3-micron-diameter dioctyl phthalate (DOP) smoke at a flow rate of 85 liters per minute.

9. Each canister must have a label warning that gas masks should be used only in atmospheres containing sufficient oxygen to support life (at least 16 percent by volume), because gas mask canisters are only designed to neutralize or remove contaminants from the air.

10. Each gas mask canister must be a distinctive color or combination of colors as indicated in the following Table. All colors used will be such that they are clearly identifiable by the user and clearly distinguishable from one another. The color coding used must offer a high degree of resistance to chipping, scaling, peeling, blistering, fading, and the effects of the ordinary atmospheres to which they may be exposed under normal conditions of storage and use. Appropriately colored pressure sensitive tape may be used for the stripes.

11. **GAS MASK TABLE**

Atmospheric Contaminants To Be Protected Against	Colors Assigned*
Acid gases	White.
Hydrocyanic acid gas	White with 1/2 inch green stripe completely around the canister near the bottom.
Chlorine gas	White with 1/2 inch yellow stripe completely around the canister near the bottom.
Organic vapors	Black.
Ammonia gas	Green.
Acid gases and ammonia gas	Green with 1/2 inch white stripe completely around the canister near the bottom.
Carbon monoxide	Blue.
Acid gases and organic vapors	Yellow.
Hydrocyanic acid gas and chloropicrin vapors	Yellow with 1/2 inch blue stripe completely around the canister near the bottom.
Acid gases, organic vapors and ammonia gases	Brown.

Atmospheric Contaminants To Be Protected Against	*Colors Assigned**
Radioactive materials, excepting tritium and noble gases	Purple (Magenta).
Particulates (dusts, fumes, mists, fogs, or smokes) in combination with any of the above gases or vapors	Canister color for contaminant, as designated above, with ½ inch gray stripe completely around the near the top.
All of the above atmospheric contaminants	Red with 1/2 inch gray stripe completely around the canister near the top.

* Gray will not be assigned as the main color for a canister designed to remove acids or vapors.

> NOTE: Orange will be used as a complete body, or stripe color to represent gases not included in this table. The user will need to refer to the canister label to determine the degree of protection the canister will afford.

12. A record must be kept showing the calendar date of the inspection of each emergency-use respirator and the results of each such inspection.

VIII. PROGRAM EVALUATION

1. The continued effectiveness of this respiratory protection program will be regularly evaluated by the person designated to be responsible for this respiratory protection program (*See* Section II above).

2. Inspections to determine compliance with its requirements will be conducted periodically.

RESPIRATOR TRAINING RECORD

Group/Dept./Craft: _____   Date: _____

Project Location: _____

Respirator Training was given in accordance with the Respiratory Protection Program. Quantitive/Qualitative [indicate which] fit testing was completed using appropriate test procedures. Respirators were issued to the following individuals and each individual checked for proper fitting.

Employee's Signature	Respirator Style & No.	Cartridge Type & No.	Fit Tested For Exposure To	Certified Fit: Yes	Certified Fit: No	Trainer's Initials

Complete All Sections Fully and Submit to:

_____

_____
Trainer's Signature

_____
Supervisor's Signature

APPENDIX A
29 CFR 1910.134 —RESPIRATORY PROTECTION

§ 1910.134 Respiratory Protection

(a) Permissible practice. (1) In the control of those occupational diseases caused by breathing air contaminated with harmful dusts, fogs, fumes, mists, gases, smokes, sprays, or vapors, the primary objective shall be to prevent atmospheric contamination. This shall be accomplished as far as feasible by accepted engineering control measures (for example, enclosure or confinement of the operation, general and local ventilation, and substitution of less toxic materials). When effective engineering controls are not feasible, or while they are being instituted, appropriate respirators shall be used pursuant to the following requirements.

(2) Respirators shall be provided by the employer when such equipment is necessary to protect the health of the employee. The employer shall provide the respirators which are applicable and suitable for the purpose intended. The employer shall be responsible for the establishment and maintenance of a respiratory protective program which shall include the requirements outlined in paragraph (b) of this section.

(3) The employee shall use the provided respiratory protection in accordance with instructions and training received.

(b) Requirements for a minimal acceptable program.

(1) Written standard operating procedures governing the selection and use of respirators shall be established.

(2) Respirators shall be selected on the basis of hazards to which the worker is exposed.

(3) The user shall be instructed and trained in the proper use of respirators and their limitations.

(4) [Reserved]

(5) Respirators shall be regularly cleaned and disinfected. Those used by more than one worker shall be thoroughly cleaned and disinfected after each use.

(6) Respirators shall be stored in a convenient, clean, and sanitary location.

(7) Respirators used routinely shall be inspected during cleaning. Worn or deteriorated parts shall be replaced. Respirators for emergency use such as self-contained devices shall be thoroughly inspected at least once a month and after each use.

(8) Appropriate surveillance of work area conditions and degree of employee exposure or stress shall be maintained.

(9) There shall be regular inspection and evaluation to determine the continued effectiveness of the program.

(10) Persons should not be assigned to tasks requiring use of respirators unless it has been determined that they are physically able to perform the work and use the equipment. The local physician shall determine what health and physical conditions are pertinent. The respirator user's medical status should be reviewed periodically (for instance, annually).

(11) Respirators shall be selected from among those jointly approved by the Mine Safety and Health Administration and the National Institute for Occupational Safety and Health under the provisions of 30 CFR part 11.

(c) Selection of respirators. Proper selection of respirators shall be made according to the guidance of American National Standard Practices for Respiratory Protection Z88.2-1969.

(d) Air quality. (1) Compressed air, compressed oxygen, liquid air, and liquid oxygen used for respiration shall be of high purity. Oxygen shall meet the requirements of the United States Pharmacopoeia for medical or breathing oxygen. Breathing air shall meet at least the requirements of the specification for Grade D breathing air as described in Compressed Gas Association Commodity Specification G-7.1-1966. Compressed oxygen shall not be used in supplied-air respirators or in open circuit self-contained breathing apparatus that have previously used compressed air. Oxygen must never be used with air line respirators.

(2) Breathing air may be supplied to respirators from cylinders or air compressors.

(i) Cylinders shall be tested and maintained as prescribed in the Shipping Container Specification Regulations of the Department of Transportation (49 CFR part 178).

(ii) The compressor for supplying air shall be equipped with necessary safety and standby devices. A breathing air-type compressor shall be used. Compressors shall be constructed and situated so as to avoid entry of contaminated air into the system and suitable in-line air purifying sorbent beds and filters installed to further assure breathing air quality. A receiver of sufficient capacity to enable the respirator wearer to escape from a contaminated atmosphere in event of compressor failure, and alarms to indicate compressor failure and overheating shall be installed in the system. If an oil-lubricated compressor is used, it shall have a high-temperature or carbon monoxide alarm, or both. If only a high-temperature alarm is used, the air from the compressor shall be frequently tested for carbon monoxide to insure that it meets the specifications in paragraph (d)(1) of this section.

(3) Air line couplings shall be incompatible with outlets for other gas systems to prevent inadvertent servicing of air line respirators with nonrespirable gases or oxygen.

(4) Breathing gas containers shall be marked in accordance with American National Standard Method of Marking Portable Compressed Gas Containers to Identify the Material Contained, Z48.1-1954; Federal Specification BB-A-1034a, June 21, 1968, Air, Compressed for Breathing Purposes; or Interim Federal Specification GG-B-00675b, April 27, 1965, Breathing Apparatus, Self-Contained.

(e) Use of respirators. (1) Standard procedures shall be developed for respirator use. These should include all information and guidance necessary for their proper selection, use, and care. Possible emergency and routine uses of respirators should be anticipated and planned for.

(2) The correct respirator shall be specified for each job. The respirator type is usually specified in the work procedures by a qualified individual supervising the respiratory protective program. The individual issuing them shall be adequately instructed to insure that the correct respirator is issued.

(3) Written procedures shall be prepared covering safe use of respirators in dangerous atmospheres that might be encountered in normal operations or in emergencies. Personnel shall be familiar with these procedures and the available respirators.

(i) In areas where the wearer, with failure of the respirator, could be overcome by a toxic or oxygen-deficient atmosphere, at least one additional man shall be present. Communications (visual, voice, or signal line) shall be maintained between both or all individuals present. Planning shall be such that one individual will be unaffected by any likely incident and have the proper rescue equipment to be able to assist the other(s) in case of emergency.

(ii) When self-contained breathing apparatus or hose masks with blowers are used in atmospheres immediately dangerous to life or health, standby men must be present with suitable rescue equipment.

(iii) Persons using air line respirators in atmospheres immediately hazardous to life or health shall be equipped with safety harnesses and safety lines for lifting or removing persons from hazardous atmospheres or other and equivalent provisions for the rescue of persons from hazardous atmospheres shall be used. A standby man or men with suitable self-contained breathing apparatus shall be at the nearest fresh air base for emergency rescue.

(4) Respiratory protection is no better than the respirator in use, even though it is worn conscientiously. Frequent random inspections shall be conducted by a qualified individual to assure that respirators are properly selected, used, cleaned, and maintained.

(5) For safe use of any respirator, it is essential that the user be properly instructed in its selection, use, and maintenance. Both supervisors and workers shall be so instructed by competent persons. Training shall provide the men an opportunity to handle the respirator, have it fitted properly, test its face-piece-to-face seal, wear it in normal air for a long familiarity period, and, finally, to wear it in a test atmosphere.

(6) Every respirator wearer shall receive fitting instructions including demonstrations and practice in how the respirator should be worn, how to adjust it, and how to determine if it fits properly. Respirators shall not be worn when conditions prevent a good face seal. Such conditions may be a growth of beard, sideburns, a skull cap that projects under the facepiece, or temple pieces on glasses. Also, the absence of one or both dentures can seriously affect the fit of a facepiece. The worker's diligence in observing these factors shall be evaluated by periodic

check. To assure proper protection, the facepiece fit shall be checked by the wearer each time he puts on the respirator. This may be done by following the manufacturer's facepiece fitting instructions.

(ii) Providing respiratory protection for individuals wearing corrective glasses is a serious problem. A proper seal cannot be established if the temple bars of eye glasses extend through the sealing edge of the full facepiece. As a temporary measure, glasses with short temple bars or without temple bars may be taped to the wearer's head. Wearing of contact lenses in contaminated atmospheres with a respirator shall not be allowed. Systems have been developed for mounting corrective lenses inside full facepieces. When a workman must wear corrective lenses as part of the facepiece, the facepiece and lenses shall be fitted by qualified individuals to provide good vision, comfort, and a gas-tight seal.

(iii) If corrective spectacles or goggles are required, they shall be worn so as not to affect the fit of the facepiece. Proper selection of equipment will minimize or avoid this problem.

(f) Maintenance and care of respirators. (1) A program for maintenance and care of respirators shall be adjusted to the type of plant, working conditions, and hazards involved, and shall include the following basic services:

(i) Inspection for defects (including a leak check),(ii) Cleaning and disinfecting,(iii) Repair,(iv) Storage Equipment shall be properly maintained to retain its original effectiveness.(2) (i) All respirators shall be inspected routinely before and after each use. A respirator that is not routinely used but is kept ready for emergency use shall be inspected after each use and at least monthly to assure that it is in satisfactory working condition.

(ii) Self-contained breathing apparatus shall be inspected monthly. Air and oxygen cylinders shall be fully charged according to the manufacturer's instructions. It shall be determined that the regulator and warning devices function properly.

(iii) Respirator inspection shall include a check of the tightness of connections and the condition of the facepiece, headbands, valves, connecting tube, and canisters. Rubber or elastomer parts shall be inspected for pliability and signs of deterioration. Stretching and manipulating rubber or elastomer parts with a massaging action will keep them pliable and flexible and prevent them from taking a set during storage.

(iv) A record shall be kept of inspection dates and findings for respirators maintained for emergency use.

(3) Routinely used respirators shall be collected, cleaned, and disinfected as frequently as necessary to insure that proper protection is provided for the wearer. Respirators maintained for emergency use shall be cleaned and disinfected after each use.

(4) Replacement or repairs shall be done only by experienced persons with parts designed for the respirator. No attempt shall be made to replace components or to make adjustment or repairs beyond the manufacturer's recommendations. Reducing or admission valves or regulators shall be returned to the manufacturer or to a trained technician for adjustment or repair.

(5) (i) After inspection, cleaning, and necessary repair, respirators shall be stored to protect against dust, sunlight, heat, extreme cold, excessive moisture, or damaging chemicals. Respirators placed at stations and work areas for emergency use should be quickly accessible at all times and should be stored in compartments built for the purpose. The compartments should be clearly marked. Routinely used respirators, such as dust respirators, may be placed in plastic bags. Respirators should not be stored in such places as lockers or tool boxes unless they are in carrying cases or cartons.

(ii) Respirators should be packed or stored so that the facepiece and exhalation valve will rest in a normal position and function will not be impaired by the elastomer setting in an abnormal position.

(iii) Instructions for proper storage of emergency respirators, such as gas masks and self-contained breathing apparatus, are found in "use and care" instructions usually mounted inside the carrying case lid.

(g) Identification of gas mask canisters.

(1) The primary means of identifying a gas mask canister shall be by means of properly worded labels. The secondary means of identifying a gas mask canister shall be by a color code.

(2) All who issue or use gas masks falling within the scope of this section shall see that all gas mask canisters purchased or used by them are properly labeled and colored in accordance with these requirements before they are placed in service and that the labels and colors are properly maintained at all times thereafter until the canisters have completely served their purpose.

(3) On each canister shall appear in bold letters the following:

(i)- Canister for

(Name for atmospheric contaminant)

or

I23 Type N Gas Mask Canister

(ii) In addition, essentially the following wording shall appear beneath the appropriate phrase on the canister label: "For respiratory protection in atmospheres containing not more than_____percent by volume of _____."

(Name of atmospheric contaminant)

(4) Canisters having a special high- efficiency filter for protection against radionuclides and other highly toxicparticulates shall be labeled with a statement of the type and degree of protection afforded by the filter. The label shall be affixed to the neck end of, or to the gray stripe which is around and near the top of, the canister. The degree of protection shall be marked as the percent of penetration of the canister by a 0.3-micron-diameter dioctyl phthalate (DOP) smoke at a flow rate of 85 liters per minute.

(5) Each canister shall have a label warning that gas masks should be used only in atmospheres containing sufficient oxygen to support life (at least 16 percent by volume), since gas mask canisters are only designed to neutralize or remove contaminants from the air.

(6) Each gas mask canister shall be painted a distinctive color or combination of colors indicated in Table I-1. All colors used shall be such that they are clearly identifiable by the user and clearly distinguishable from one another. The color coating used shall offer a high degree of resistance to chipping, scaling, peeling, blistering, fading, and the effects of the ordinary atmospheres to which they may be exposed under normal conditions of storage and use. Appropriately colored pressure sensitive tape may be used for the stripes.

Table I-1

Atmospheric Contaminants To Be Protected Against	Colors Assigned[1]
Acid gases	White.
Hydrocyanic acid gas	White with \1/2\-inch green stripe completely around the canister near the bottom.
Chlorine gas	White with \1/2\-inch yellow stripe completely around the canister near the bottom.
Organic vapors	Black.
Ammonia gas	Green.
Acid gases and ammonia gas	Green with \1/2\-inch white stripe completely around the canister near the bottom.
Carbon monoxide	Blue.
Acid gases and organic vapors	Yellow.
Hydrocyanic acid gas and chloropicrin	Yellow with \1/2\-inch blue stripe completely around the canister near the bottom.
Acid gases, organic vapors, and ammonia gas	Brown.
Radioactive materials, excepting tritium and noble gases.	Purple (Magenta).
Particulates (dusts, fumes, mists, contaminant, as fogs, or smokes) in combination with any of the above gases or vapors	Canister color for designated with 1/2\-inch gray stripe completely around the canister near the top.
All of the above atmospheric contaminants	Red with \1/2\-inch gray stripe completely around the canister near the top.

[1] Gray will not be assigned as the main color for a canister designed to remove acids or vapors. Note: Orange will be used as a complete body, or stripe color to represent gases not included in this table. The user will need to refer to the canister label to determine the degree of protection the canister will afford.

(Approved by the Office of Management and Budget under control number 1218-0099)

[39 FR 23502, June 27, 1974, as amended at 43 FR 49748, Oct. 24, 1978; 49 FR 5322, Feb. 10, 1984; 49 FR 18295, Apr. 30, 1984; 58 FR 35309, June 30, 1993]

APPENDIX B
OSHA INSTRUCTION CPL 2-2.54
RESPIRATORY PROTECTION
PROGRAM MANUAL

U.S. Department of Labor Assistant Secretary for
Occupational Safety and Health
Washington, D.C. 20210

OSHA Instruction CPL 2-2.54
FEB 10 1992
Office of Science and Technology Assessment

Subject: **Respiratory Protection Program Manual**

A. **Purpose.** This instruction sets forth accepted practices for respirator users, provides information and guidance on the proper selection, use, and care of respirators, and contains requirements for establishing an OSHA respirator program.

B. **Scope.** This instruction applies to all OSHA employees who needs to wear a respirator to perform her/his duties.

C. **Implementation.** Detailed instructions for implementing the above requirements are prescribed in the following chapters.

D. **Background.**

   1. Occupational Safety and Health Administration (OSHA) Compliance Safety and Health Officers (CSHO's) as well as other Agency personnel may be exposed to a variety of respiratory hazards while conducting safety and health compliance inspections, consultation or monitoring visits.

   2. The human respiratory system can be protected by avoiding or minimizing exposure to harmful substances; however, in some cases this may not be possible and an appropriate respirator shall be required. Certain respirators can reduce/remove many contaminants from an atmosphere. When concentrations of these contaminants are too high to be reduced/removed or when oxygen levels are too low, other respirators are available which can supply breathing quality air to the wearer. Therefore, proper selection of the appropriate respirator for the conditions at hand is mandatory.

E. **Federal Program Change.** This instruction describes a Federal program change which affects State programs. Each Regional Administrator shall:

   1. Ensure that this instruction is forwarded to each State designee using a format consistent with the Plan Change Two-Way Memorandum in Appendix P, OSHA Instruction STP 2.22A, CH-2.

OSHA Instruction 2-2.54
FEB 10 1992
Office of Science and Technology Assessment

OCCUPATIONAL SAFETY AND HEALTH ADMINISTRATION

RESPIRATORY PROTECTION PROGRAM MANUAL

Office of Science and Technology Assessment

1991

OSHA Instruction CPL 2-2.54
FEB 10 1992
Office of Science and Technology Assessment

TABLE OF CONTENTS

Chapter I. **RESPONSIBILITIES**

    A. NATIONAL OFFICE 34
    B. REGIONAL OFFICE 34
    C. AREA OFFICE 35
    D. PROGRAM COORDINATOR 36
    E. CSHO 37

Chapter II. **RESPIRATORY PROTECTION PROGRAM ADMINISTRATION**   39

Chapter III. **SELECTION**

    A. GENERAL 43
    B. AIR-PURIFYING RESPIRATORS 44
    C. ATMOSPHERE SUPPLYING RESPIRATORS 46
    D. EMERGENCY ESCAPE RESPIRATORS 48

Chapter IV. **TRAINING** 51

Chapter V. **FIT TESTING**

    A. QUANTITATIVE FIT TESTING 52

        1. PURPOSE 52
        2. DEFINITIONS 52
        3. APPARATUS 52
        4. PROCEDURAL REQUIREMENTS 54
        5. EXERCISE REGIMEN 56
        6. CALCULATION OF FIT FACTORS 58

    B. QUALITATIVE FIT TESTING

        1. PURPOSE 58
        2. RESPIRATOR SELECTION 59
        3. FIT TEST PROCEDURE 60

    C. OTHER REQUIREMENTS 62
    D. RECORDKEEPING 64

Chapter VI. **MAINTENANCE** 65

OSHA Instruction CPL 2-2.54
FEB 1 0 1992
Office of Science and Technology Assessment

Appendix A Checklist and Inspection form for MSA SCBA   67

Appendix B Inspection and Maintenance Procedures for Survivair SCBA   72

OSHA Instruction 2-2.54
FEB 10 1992
Office of Science and Technology Assessment

Chapter I

RESPONSIBILITIES

The guidelines established in this program may be supplemented as local needs dictate. However, the direction herein MUST be included in Regional and Area Office programs.

A. **National Office - Directorate of Technical Support shall**.

   1. Issue guidelines and directives that initiate and update the program.

   2. Assist the Regional Administrators in complying with the programs.

   3. Recommend appropriate respiratory protective equipment.

   4. Audit and review the effectiveness of the program.

   5. Administrate the respiratory protection program for the National Office personnel.

B. **Regional Office/Regional Administrator shall**.

   1. Establish a Regional respirator program and ensure that each Area Office establishes/implements a consistent program.

   2. Examine and evaluate the Area Office's program to determine its effectiveness.

   3. Recommend systems for complying with the program and assist in technical requirements.

   4. Maintain models and sizes of air-purifying type of respirator available for selection and fitting. These models and sizes will be representative of respirators from major manufacturers.

   5. Ensure the purchase of the proper type of equipment in adequate quantities.

   6. Establish a respirator program for Regional Office staff as outlined in Area Office requirements of this section.

OSHA Instruction CPL 2-2.54
FEB 10 1992
Office of Science and Technology Assessment

7. Appoint a Regional coordinator to administer and evaluate the overall program. The Regional coordinator monitors field office adherence to the procedures established in this directive. A written report detailing program effectiveness measures shall be submitted to the Director, Directorate of Technical Support annually.

8. <u>ANNUAL REPORT FORMAT</u>. The following information shall be included in the annual report.

    a. Listing by field office by name of those quantitatively fit tested including:

        (1) Current test date, quantitative fit testing /(QNFT) method (Portacount or Photometric), respirator models evaluated and fit factors (FF) derived.

        (2) Most recent previous quantitative fit test date, model selected and fit factors (FF) obtained.

    b. Resources used to administer the program including:

        (1) Staff-years to perform quantitative fit testing including subject time.

        (2) Staff-years to administer program which includes training, evaluation and report writing.

    c. Comments and suggestions concerning program administration, effectiveness, technical problems relating to equipment and other relevant issues.

C. <u>Area Office/Area Director shall</u>.

1. Assume responsibility for administrating the program in the Area Office.

2. Ensure that the respirator program is adhered to by the CSHOs.

OSHA Instruction CPL 2-2.54
FEB 1 0 1992
Office of Science and Technology Assessment

3. Delegate authority for the coordination of the program, or parts of the program, to a person(s) trained in the use and care of specific types of equipment as determined by the Area Director. Responsibility and authority for the respirator program shall be assigned to a single person. This individual will be designated the Respiratory Protection Program Coordinator.

4. Provide an evaluation of the respiratory protection program to the Regional Coordinator annually. The report shall include the elements listed under **THE ANNUAL REPORT FORMAT** (b.8.) of this chapter.

D. <u>Program Coordinator (Regional and Area Offices) shall</u>:

1. Attend the OSHA Training Institute respiratory protection course.

2. Develop standard operating procedures (SOPs).

3. Be responsible for cleaning, maintenance and storage of all respirators not routinely used, or not individually assigned.

4. Maintain respirator supplies, including spare parts; obtain new equipment and maintain non-individually assigned equipment ready for reissue.

5. Ensure that sufficient quantities of filters and chemical cartridges and canisters for specific contaminants shall be available in each Area Office.

6. Aid CSHOs in respirator fit testing. Each CSHO shall receive respirator fitting instructions and undergo at least annual quantitative fit testing to select the best fitting facepiece. Fit testing shall be performed more frequently to meet the requirements prescribed in specific standards such as asbestos or acrylonitrile.

7. Ensure that each Area Office properly maintains long service life and emergency escape self-contained breathing apparatus (SCBAs). A fully charged spare cylinder shall be available for routine use SCBA. These cylinders will be properly inspected and maintained.

OSHA Instruction 2-2.54
FEB 1 0 1992
Office of Science and Technology Assessment

8. Provide additional training and information for CSHOs in the correct use, maintenance, cleaning and care of respirators. Respirators shall be repaired under the direction of the PROGRAM COORDINATOR.

9. Evaluate periodically the effectiveness of the respirator program. The following elements should be considered when evaluating the program's effectiveness:

   a. The proper types of respirators are selected.

   b. The wearers are properly trained.

   c. The correct respirators are issued.

   d. The respirators are worn properly.

   e. The respirators are properly maintained and cleaned.

   f. The respirators are properly stored.

   g. Fit testing is conducted properly.

   h. All pertinent records are kept.

   i. Submit a report to the Regional Coordinator after each evaluation of the program. The report shall include the results of inspection, the respirator program administration, investigating wearer acceptance, any inadequacy of the program and any action taken to correct the deficiency, and target dates for planning implementation. The Regional Coordinator shall prepare a summary of each area office's program to the Regional Administrator.

E. **Compliance Safety and Health Officers.**

   1. Individuals assigned tasks which require respiratory protective equipment will use the appropriate equipment in accordance with this instruction.

   2. Each CSHO shall clean, disinfect and properly store as necessary, the respirator assigned for personal use. Cleaning agents shall be available in each Area Office.

OSHA Instruction 2-2.54
FEB 1 0 1992
Office of Science and Technology Assessment

3. Each CSHO shall inspect the respirator before each use and after cleaning and disinfecting. The inspection shall include a check for defects, missing parts and a facepiece leak check. If a respirator is found defective, it shall be returned to the PROGRAM COORDINATOR for repair.

4. Each CSHO shall comply with fit test requirements and all other provisions of this directive.

5. Each CSHO shall attend the OSHA Training Institute respirator course.

F. **National Office Personnel.**

1. Individuals assigned tasks which require respiratory protective equipment shall use the appropriate equipment in accordance with this instruction.

2. Each respirator wearer shall clean, disinfect and properly store as necessary, the respirator assigned for personal use.

3. Each respirator wearer shall inspect the respirator before each use and after cleaning and disinfecting. The inspection shall include a check for defects, missing parts and a facepiece leak check. If a respirator is found defective, it shall be returned to the Directorate of Technical Support for repair.

4. Each respirator wearer shall comply with fit test requirements and all other provisions of this directive.

5. Each respirator wearer shall attend the OSHA Training Institute respirator course.

OSHA Instruction CPL 2-2.54
FEB 10 1992
Office of Science and Technology Assessment

Chapter II

RESPIRATORY PROTECTION PROGRAM ADMINISTRATION

A. Respirators and accessories shall be available for OSHA employees at job sites. Respirators shall be worn whenever requested by the employer, as well as during the time an OSHA employee is in a contaminated area performing air sampling, since the possibility of overexposure exists. Even in the event that air sampling is not being performed, and an OSHA employee is in a contaminated area which is likely to exceed the PEL (e.g., a previous citation for overexposure was issued to the company for that area), respirators shall also be worn. Exceptions to this policy may include air samples taken for screening purposes or other situations as individually approved by the supervisor. OSHA personnel are encouraged to wear respirators at any time they feel it is appropriate for their self-protection.

B. Written standard operating procedures (SOP) shall be prepared including all information and guidance necessary for proper respirator selection, use, care, and maintenance. The written SOP shall be provided by the Regional Office and shall be modified to suit the needs of each field location.

C. Respirators shall be selected on the basis of hazards to which the person is exposed with consideration given to both safety and health factors as well as probable risk. Individuals issuing respirators shall be adequately instructed to ensure that the correct respirator is issued and that each respirator is complete. To the extent possible half-mask respirators should be assigned to individual workers for their exclusive use.

D. Before initial use, all new respirators shall be washed, cleaned, sanitized and inspected per respirator manufacturer's instruction. The oxygen content of the SCBA cylinder shall be verified. Each respirator shall be properly fitted and a leakage test performed. Before each use, both positive and negative pressure fit checks shall be conducted. The user shall be instructed and trained in the proper use of respirators and informed about their limitations.

OSHA Instruction CPL 2-2.54
FEB 10 1992
Office of Science and Technology Assessment

E. Respirators shall be cleaned and disinfected by the wearer after use. Those used by more than one CSHO shall be thoroughly cleaned and disinfected after each use. Respirators shall be stored in a convenient, clean, and sanitary location free of contaminants which may damage the components of a respirator.

F. Respirators used on a regular basis shall be inspected during cleaning. Trained personnel shall replace worn or deteriorated parts with parts designed for the respirator. No attempt shall be made to replace components or to make adjustments or repairs beyond the manufacturer's recommendations. Self-contained breathing apparatus (SCBA) and emergency escape SCBA shall be thoroughly inspected at least once a month and after each use, and a written record kept of inspection dates and findings. Since the SCBA is complex equipment, service of the device shall be limited to trained personnel certified by the manufacturer.

G. Supervisors and workers shall be instructed and trained in the selection, use, care, and maintenance of respiratory protective devices. Training shall provide each user an opportunity to handle the respirator, to have it fitted properly, to test its facepiece-to-face seal, to wear it in normal air for a familiarization period, and to wear it in a test atmosphere. Retraining will be performed as needed or at least annually to ensure an effective program.

H. There shall be regular inspections and evaluations to determine the continued effectiveness of the program.

I. Clean shaven skin must be in contact with all respirator sealing surfaces. Even a mild growth of whiskers may interfere with this seal. In addition, respirators shall not be worn when conditions such as sideburns, a skull cap that projects under the facepiece, temple pieces on corrective spectacles or goggles, or the absence of one or both dentures prevent a good facepiece--to-face seal. Therefore, while on duty, all OSHA employees within the scope of this policy must be clean shaven in the areas of

OSHA Instruction CPL 2-2.54
FEB 10 1992
Office of Science and Technology Assessment

the respirator face sealing surface and the face. If hair growth, other than in the clean shaven area of facepiece-to-face seal, interferes with the proper function of the respirator such as the exhalation valve, then it shall be altered or removed so as to eliminate interference. The Agency's position is to provide negative pressure, half-mask or full-face piece respirators that can be tested with available fit testing equipment. The Agency will also provide tight fitting powered air-purifying respirators (PAPRs) to CSHO's upon request.

J. Corrective lenses which interfere with the facepiece-to-face sealing area shall not be used with a full facepiece. Contact lenses may be worn with a full facepiece with the approval of the Regional Program Coordinator.

K. Single use, disposable or maintenance free respirators will not be used by OSHA personnel. Since the CSHOs may encounter different air contaminants during an inspection, air-purifying respirators with replaceable cartridges shall be used because these devices provide more flexibility and reduce the number of single respirators which need to be carried by the CSHOs. Furthermore, disposable, maintenance free or single use respirators provide a poorer facepiece seal than multi-sized elastomeric facepieces and often it is difficult to perform an effective negative and/or positive pressure facepiece leakage test. Only "mechanical type" high-efficiency particulate air (HEPA) filters enclosed in cartridges or canisters are acceptable for protection against any particulate exposure because efficiency of these filters does not change with dust loading and ambient conditions.

L. Any respirator may produce undesirable effects on the wearer. Respirators are uncomfortable, and may reduce field of vision, require the individual to carry extra weight, place an additional burden on the respiratory system, cause a feeling of claustrophobia, and may result in a general feeling of anxiety. The two areas of greatest interest as far as physiological effects are concerned are the respiratory system and the cardiovascular system.

M. Individuals shall be examined medically before being assigned to use respirators. The examining physician shall be given information about the equipment to be used. He or

OSHA Instruction CPL 2-2.54
FEB 10 1992
Office of Science and Technology Assessment

she should know whether it produces additional inspiratory and expiratory stress, whether it represents an additional weight, such as self-contained breathers, and whether it may cause an increase in the metabolic heat load, such as chemical protective clothing.

1. CSHOs shall not be assigned to tasks requiring use of respirators unless it has been determined by medical authorities that they are physically able to perform their duties while wearing the prescribed respirators and chemical protective clothing. The examining physician shall provide a written opinion which describes the ability of the CSHO to wear the prescribed respirator and recommends limitations on the use of respirators if any. The report and opinion shall be forwarded to the Office of Occupational Medicine, Directorate of Technical Support.

2. The medical status of the respirator user shall be reviewed as part of the examinations required under the agency CSHO physical program. This review shall be performed with the assistance of the Office of Occupational Medicine, Directorate of Technical Support.

OSHA Instruction CPL 2-2.54
FEB 10 1992
Office of Science and Technology Assessment

Chapter III

RESPIRATOR SELECTION

A. **General**.

1. The guidelines outlined in this section provide assistance in the selection of appropriate respiratory protection by OSHA personnel. The Agency shall provide appropriate approved respiratory protective devices and the CSHOs shall use these devices whenever necessary to protect their health due to the nature of the work environment. It is important that the safety and health professional assess the potential hazards and degree of controls which can be exercised over each situation. The respiratory protective devices selected in each situation will depend upon the information from a qualitative and/or quantitative determination of the hazard. The professional judgment is essential to insure appropriate selections of respirators.

2. The nature of respiratory hazard, as it refers to the selection and classification of respirators, depends upon the atmospheric oxygen concentration; a contaminant's physical state, toxicity, and concentration; the presence of other contaminants or stress factors in the working environment; and worker exposure time and susceptibility. Respiratory hazards may be classified as gas and vapor contaminants (immediately or not immediately dangerous to life or health), particulate contaminants (immediately or not immediately dangerous to life or health), and oxygen deficiencies. Each classification requires a different type of respiratory protection.

3. In the selection and use of respiratory protective devices, health and safety factors must be considered, such as nature of the hazard, intended uses and limitations of respiratory protective devices, movement and work rate limitations, emergency escape time and distance requirements, and training requirements.

OSHA Instruction CPL 2-2.54
FEB 10 1992
Office of Science and Technology Assessment

    4. Among additional general considerations in determining the appropriate respirator are sorbet efficiencies, odor warning properties, eye irritation potential, protection factors (PF), lower flammability limit (LFL), and conditions which are immediately dangerous to life or health (IDLH -- as defined in 1910.120). Reference materials are also available to assist in determining the general conditions or situations which would indicate the most prudent use of respirator protection (list available at the Directorate of Technical Support).

B. **Air-Purifying Respirators.**

    1. In general, air-purifying cartridge or canister respirators will be allowed if the contaminant(s) is known, the concentration(s) is known, the air-purifying element provides adequate protection for the air contaminant(s), and the contaminant(s) has good warning properties. Certain specific health standards permit the use of air-purifying respirators even though the chemical has poor or no warning properties. This type of respirator may either be equipped with chemical cartridges or a canister for protection against gases and vapors.

    2. With regard to particulate respirators, an increase in breathing resistance (comfortable breathing impaired) occurs as a result of the challenge particulate lodging on the respirator filter. Since this is a subjective indicator, the HEPA filter cartridges should be replaced at least once a week in moderate to dusty workplaces or every three weeks in low dust environments, when contamination of the cartridge surface is noticed, or when the filter has been dropped or subjected to other trauma.

    3. A much more insidious problem occurs with regard to end of service life indication for gas and vapor cartridge/canister equipped respirators. End of service life indication is generally based on an individual's ability to detect (e.g., taste, smell) the contaminant within the respirator wearer's facepiece. This guideline is totally subjective and may expose the respirator user to considerable risk.

OSHA Instruction CPL 2-2.54
FEB 10 1992
Office of Science and Technology Assessment

The basis of the subject detection principle is the assumption that gases/vapors in question have good warning properties. Often the necessary information relative to this liming factor is difficult to obtain or may not exist; i.e., the odor threshold of a particular material. Studies have also shown that certain people have only a moderately developed sense of smell. Olfactory fatigue may occur to individuals acclimatized to the odor.

Some gaseous contaminants will migrate across the adsorbent or absorbent bed while the respirator is not in use, such as overnight. This migration subjects the user to an initial dose of the contaminant when the respirator is again placed in service. Therefore, as a minimum, gas/vapor cartridges shall be disposed of after each day's activities no matter how short those activities were. A day's activities would begin when the plastic seal or bag is removed from the cartridges allowing those cartridges to be exposed to moisture. These cartridges, even if they are not exposed to a contaminated atmosphere, must be discarded. A label must be attached to the cartridge indicating the installation date.

4. Since odor threshold and olfactory fatigue vary among different individuals, the use of chemical cartridge respirators against substances with poor warning properties shall not be permitted unless its use is permitted in specific health standards. In this case, reliable information concerning service life must be available. Since some reactive chemicals cannot be effectively adsorbed by the sorbet, its use should also be restricted. A partial (not all inclusive) list of air contaminants with poor odor warning properties or short breakthrough time follows:

Acrolein, aniline, arsine, boron hydrides, bromine, carbon dioxide, carbon monoxide, carbonyls, carbon disulfide, cyanogen dimethylaniline, dimethyl sulfate, fluorine, hydrogen cyanide, hydrogen fluoride, hydrogen selenide, hydrogen sulfide, isocyanates: HDI, MDI, MIC and TDI, methanol, methyl bromide, methyl chloride, methyl iodine, nickel carbonyl, nitrocompounds: nitrobenzene, nitrogen oxides, nitroglycerine, nitromethane, ozone, phosgene, phosphine, phosphorous trichloride, stibine, sulfur chloride, and vinyl chloride.

OSHA Instruction 2-2.54
FEB 10 1992
Office of Science and Technology Assessment

5. With regard to air-purifying respirators, the fit factor obtained during fit testing has little predictive ability for determining the specific level of protection that will be achieved all of the time in the workplace. Because of this, the respirator assigned protection factors listed in Table I should not be exceeded no matter how high the fit factor during fit testing:

C. **Atmosphere Supplying Respirators**.

1. Normally a Compliance Safety and Health Officer, both health and safety will not enter, without prior Regional approval, an area where an atmosphere supplying respirator is required.

    Whenever possible, evaluation methods, i.e., sampling strategies, will be applied which will not require entry into an extremely hazardous area. In those situations where entry must be made into potentially oxygen deficient environments or contaminated atmospheres which are immediately dangerous to life or health, an appropriate approved atmosphere supplying respirator must be used. Some situations which may require the use of such respiratory protective equipment include entry into confined spaces, hazardous substance spill/waste disposal areas, employer requirement for specific atmosphere supplying respiratory protection and emergency investigations requiring entry into potentially IDLH atmospheres. The Regional Administrator may delegate to the Area Director the approval authority for use of SCBA in non IDLH situations.

2. An SCBA must not be used in IDLH or potentially IDLH atmospheres unless a second or a standby CSHO is present and also equipped with a SCBA. The CSHO must be in communication (visual, vocal or signal) with the other CSHO at all times. The second CSHO will be present to provide any needed assistance or rescue. A fully charged spare air cylinder must always be available. Additionally, consideration for other than

OSHA Instruction CPL 2-2.54
FEB 10 1992
Office of Science and Technology Assessment

respiratory protection needs must be given in situations where skin absorption/irritation potential may be present. The standby CSHO shall have the necessary training and equipment to perform a rescue if needed.

3. Learning and "hands on" experience with SCBA must be ongoing and continuous. Compliance Safety and Health Officers who are designated to use SCBA are to be trained in the operation and use of the devices and are to follow the manufacturers' directions and the Regional/Area Office respiratory protection program. The training must be on a regular basis (i.e., annually) and include the actual wearing and use of SCBA during exercise situations (e.g., walking). All SCBA wearers must be trained and certified by OSHA recognized training facilities. A list is available from Directorate of Technical Support.

4. The Assistant Regional Administrator for Technical Support (ARA-TS) will be consulted for assistance in determining the appropriateness of SCBA use in a specific situation. Any planned entry involving the use of SCBA shall be coordinated with the ARA-TS.

5. Since a respirator used non-routinely is principally used for hazardous situations or emergencies that occur only occasionally, routine inspection and proper maintenance are essential to assure that the designated degree of protection and useful service time are provided. Review of non-routine use respirators (i.e., SCBA) should be consistent with the recommendations of the manufacturer, the requirements of appropriate regulatory standards (e.g., 29 CFR 1910.134) and existing Regional/Area policies. Depending on the specific SCBA available for use at a location, the items included in the maintenance program may require slight modifications to encompass additional considerations as suggested by the appropriate SCBA manufacturer(s). (Refer to Appendices I and II, or user warning notices issued by NIOSH).

OSHA Instruction CPL 2-2.54
FEB 10 1992
Office of Science and Technology Assessment

D. **Emergency Escape Respirators.**

1. These devices constitute another class of non-routine use respirators. Any respirator that protects adequately against a hazardous atmosphere that has occurred suddenly may be used for escape purposes. However, these devices shall not be used for entry into this type of atmosphere even if that entry is for rescue purposes. Escape respirators shall be provided and carried by all individuals when there exists a potential for exposure to toxic materials at IDLH levels. Section F.8. of this directive shall be followed for maintenance of emergency escape respirators. Examples of these types of situations may exist in portions of refineries, chemical plants, sewage treatment plants, and hazardous waste sites etc. All emergency escape devices have limitations and these limitations must be taken into account when selecting one of these respirators. When entering a potential IDLH atmosphere, the CSHO shall assess the egress route to ensure that the emergency egress time does not exceed the capacity of the escape SCBA.

    a. If the toxic materials in question would cause eye irritation, then a full facepiece or hood must be used.

    b. Even full facepiece air-purifying emergency egress respirators such as gas masks are contaminant(s) specific and will fail to provide adequate protection at certain concentrations. In addition, those units will not provide a breathing atmosphere and therefore cannot be used in oxygen-deficient atmospheres. Since the conditions of most emergencies are unknown, OSHA emergency escape devices shall be of the atmosphere supplying variety.

    c. If it is available, only the continuous flow escape SCBA with an air flow of 70 liters per minute shall be used. If the anticipated escape time is in excess of the capacity of the continuous flow SCBA, mouthpiece SCBA shall be used. It is available on loan from the OSHA Cincinnati Laboratory (OCL).

OSHA Instruction CPL 2-2.54
FEB 1 0 1992
Office of Science and Technology Assessment

Figure III-1

RESPIRATOR SELECTION

Type	Facepiece Pressure	Maximum Use Concentration (MUC) in Multiples of PEL
Half-mask	–	10 X
Full Facepiece	–	50 X
Powered air-purifying:		
Half-mask	+	250 X
Full facepiece	+	500 X
Pressure Demand Supplied Air Respirator (PDSAR):		
Half-mask	+	250 X
Full facepiece	+	500 X
Full facepiece w/escape prov.	+	1,000 X
SCBA:		
Entry and escape:		
Full facepiece pressure	+	IDLH and unknown
Demand		Concentrations
Escape only:		
Continuous flow	+	IDLH
Mouthpiece	–	IDLH

Notes:

1. *Only a half-mask with interchangeable cartridges is acceptable.*

2. *Respirator assigned for higher concentrations may be used at lower concentrations.*

3. *Full facepiece is required if eye irritation is experienced.*

OSHA Instruction CPL 2-2.54
FEB 10 1992
Office of Science and Technology Assessment

4. *A minimum service life of 60 minutes is required for sorbet cartridges and canisters to provide adequate protection against air contaminants having poor odor warning properties. The maximum use concentration*

   (MUC) of a respirator for protection against gases or vapors is limited by the service life of the sorbet. For example, if the cartridges used for protection against compound X have only a service life of 60 minutes at a concentration of 50 times the PEL, then the MUC for a full facepiece PAPR equipped with these cartridges is only 50 times rather than 500 times the PEL listed in the respirator selection table.

5. *If it is available, compressed air having higher purity than Grade D, such as Grade H, or J should be used for the SAR and the SCBA.*

6. *Goggles supplied with the mouthpiece escape SCBA shall be used when eye irritation is experienced.*

7. *Pressure demand supplied air respirators equipped with an auxiliary self-contained air supply may be used under IDLH conditions when the anticipated use time is longer than the service life of the SCBA and the auxiliary air cylinder provides sufficient time for escape.*

8. *If it is available, the continuous flow escape SCBA having an air flow of 70 liters per minutes shall be used.*

9. *Air-purifying respirators may not be used in oxygen deficient or IDLH atmosphere.*

10. *Tight fitting PAPRs with a minimum flow rate of 170 liters per minute shall be used.*

OSHA Instruction CPL 2-2.54
FEB 10 1992
Office of Science and Technology Assessment

Chapter IV

TRAINING

A. Selecting the respirator appropriate to a given hazard is important, but equally important is using the selected device properly. Proper use can be ensured by carefully training both supervisors and CSHOs in selection, use, fit testing and maintenance of respirators. Unless the reasons for the use of respiratory protective devices and instructions on proper use and maintenance are thoroughly understood and ongoing training provided, the devices will not be used or may not work properly. Minimum training activities shall include:

1. Instruction in the nature of the hazard and a discussion of what the results may be if the respirator is not used.

2. A discussion of why a certain type of respirator is used in a particular environment. The purpose of using respirators must be presented as well as a description of respirator capabilities and limitations.

3. Periodic instruction and training in actual respirator use including fit testing. Wearers of SCBAs and emergency escape devices must be retrained every year. This training will include actual full service time operations of the unit. End of service life indicator recognition will also be covered.

4. Recognition of emergency situations and methods to deal with such situations must be covered. Cleaning and maintenance of respirators will also be covered.

5. All supervisors and CSHOs shall attend the OSHA Training Institute course on Respiratory Protection.

OSHA Instruction CPL 2-2.54
FEB 10 1992
Office of Science and Technology Assessment

Chapter V

FIT TESTING

A. **General**.

There are two methods available for fit testing respirators, quantitative fit testing (QNFT) and qualitative fit testing (QLFT). QNFT shall be used for fit testing all CSHOs. QLFT may be used to meet the requirements for a specific standard such as asbestos or acrylonitrile.

The qualitative and quantitative fit test procedures in this section shall be incorporated into all Regional and Area Office programs.

B. **Quantitative Respirator Fit Test (QNFT)**.

1. **Purpose** - Quantitative fit testing is performed to select the best fitting respirator for CSHOs. Fit factors far exceeding practical field use needs are generally obtained from QNFT. The fit factor is only significant in terms of selecting the respirator providing the highest factor.

2. **Definitions**.

    a. "Quantitative Fit Test" means the use of instrumentation and appropriate procedures to measure respirator effectiveness. Effectiveness refers to the respirator's ability to conform to the wearers face minimizing face to seal leakage.

    b. "Challenge Agent" means the testing agent introduced into a test chamber for photometric based systems. Ambient air is the challenge agent, when employing condensation nuclei counter (CNC) devices.

3. **Apparatus**.

    Instrumentation. Two systems are currently available OSHA-wide:

    a. Photometric - Employs the generation of a known concentration of corn oil, sodium chloride or other aerosol in a test chamber.

OSHA Instruction CPL 2-2.54
FEB 10 1992
Office of Science and Technology Assessment

- b. Condensation Nuclei Counter (CNC) - Ambient air serves as the challenge agent, no test chamber is necessary.

- c. **General Requirements.**

    (1) Test chamber. The test chamber shall be large enough to permit all test subjects to freely perform all required exercises without distributing the challenge agent concentration or the measurement apparatus. The test chamber shall be equipped and constructed so that the challenge agent is effectively isolated from the ambient air yet uniform in concentration throughout the chamber. When ambient air is used as the challenge agent, quantitative fit testing shall be conducted in an area relatively free of air contaminants (including tobacco products). The minimum ambient air particle count shall be 10,000 per c.c. before testing begins.

    (2) When testing air-purifying respirators, the normal filter or cartridge element shall be replaced with a high-efficiency particular air filter supplied by the same manufacturer.

    (3) When applicable, the sampling instrument shall be selected so that a strip chart record may be made of the test showing the rise and fall of challenge agent concentration with each inspiration and expiration at fit factors of at least 5,000. When utilizing systems that perform automatic programmed calculations, the strip chart must record the calculated average fit factor for each exercise period.

    (4) The combination of substitute air-purifying elements (if any), challenge agent, and challenge agent concentration in the test chamber shall be such that the test subject is not exposed in excess of the permissible exposure limit to the challenge agent at any time during the testing process.

OSHA Instruction CPL 2-2.54
FEB 1 0 1992
Office of Science and Technology Assessment

(5) The sampling port on the probed specimens respirator shall be placed and constructed so that there is no detectable leak around the port. A free air flow is allowed into the sampling line at all times and so there is no interference with the fit or performance of the respirator.

(6) The test chamber and test set-up shall permit the person administering the test to observe one test subject inside the chamber during the test.

(7) The equipment generating the challenge atmosphere shall maintain the concentration of challenge agent constant within a 10 percent variation for the duration of the test.

(8) The time lag (interval between an event and its being recorded on the strip chart) of the instrumentation may not exceed 2 seconds.

(9) The tubing for the test chamber atmosphere and for the respirator sampling port shall be the same diameter, length and material. It shall be kept as short as possible. The smallest diameter tubing recommended by the manufacturer shall be used.

(10) The exhaust flow from the test chamber shall pass through a high-efficiency filter before release to the room.

(11) When sodium chloride aerosol is used, the relative humidity inside the test chamber shall not exceed 50 percent.

4. **Procedural Requirements.**

   a. The fitting of half-mask respirators should be started with those having multiple sizes and a variety of interchangeable cartridges and canisters such as the MSA Comfo elite-M, North-M, Survivair-M. Use either of the tests outlined below to assure that the facepiece is promptly adjusted.

OSHA Instruction CPL 2-2.54
FEB 10 1992
Office of Science and Technology Assessment

(1) Positive pressure test. With the exhaust port(s) blocked, the positive pressure upon slight exhalation should remain constant for several seconds.

(2) Negative pressure test. With the intake port(s) blocked, the negative pressure upon slight inhalation should remain constant for several seconds.

(3) After a facepiece is adjusted, the test subject shall wear the facepiece for at least 5 minutes before conducting an abbreviated qualitative fit test by using either of the methods described below and using the exercises a., b., c. and d. described in C.5. below. If the test subject wears eye glasses or safety glasses when performing his/her duties, the fit test must be performed when these glasses are worn.

   (a) Isoamyl acetate test. When using organic vapor cartridges, the test subject who is capable of smelling the odor should be unable to detect the odor of isoamyl acetate in the air near the most vulnerable portion of the facepiece seal such as the nose bridge. A combination cartridge or canister with organic vapor and high-efficiency filters shall be used when available for the particular mask being tested. The test subject shall be given an opportunity to smell the odor of isoamyl acetate before the test is conducted in a location which is separated from the test area.

   (b) Irritant fume test. When using high-efficiency filters, the test subject should be unable to detect the irritant fume (stannic chloride ventilation smoke tubes) squirted into the air near the most vulnerable portions of the facepiece seal. The test subject shall be instructed to close her/his eyes during the test period. Unless the test

OSHA Instruction CPL 2-2.54
FEB 10 1992
Office of Science and Technology Assessment

subject cannot smell the odor of isoamyl acetate, the irritant fume test should not be performed due to the irritant nature of the fume.

(c) A QNFT may be conducted only when the test subject has obtained a satisfactory fit on either of the tests above.

The following three paragraphs apply only to QNFT utilizing photometric technology.

(d) Before the subject enters the test chamber, a reasonably stable challenge agent concentration shall be obtained in the test chamber.

(e) Immediately after the subject enters the test chamber, the challenge agent concentration inside the respirator shall be measured to ensure that the peak penetration does not exceed 5 percent for a half-mask and 1 percent for a full facepiece.

(f) A stable challenge agent concentration shall be established prior to the actual start of testing.

5. **Exercise Regime**.

Prior to entering the test chamber, the test subject shall be given complete instructions as to her/his part in the test procedures. The test subject shall perform the following exercises, in the order given, for each independent test.

a. Normal Breathing (NB). In the normal standing position, without talking, the subject shall breathe normally for at least one minute.

b. Deep Breathing (DB). In the normal standing position the subject shall do deep breathing for at least one minute pausing so as not to hyperventilate.

OSHA Instruction CPL 2-2.54
FEB 10 1992
Office of Science and Technology Assessment

  c. <u>Turning head side to side (SS)</u>. Standing in place the subject shall slowly turn his/her head from side between the extreme positions to each side. The head shall be held at each extreme position for at least five seconds. Perform for at least one minute.

  d. <u>Moving head up and down (UD)</u>. Standing in place, the subject shall slowly move his/her head up and down between the extreme position straight up and the extreme position straight down. The head shall be held at each extreme position for at least five seconds. Perform for at least one minute.

  e. <u>Reading (R)</u>. The subject shall read out slowly and loud so as to be heard clearly by the test conductor or monitor. The test subject shall read the "rainbow passage".

Rainbow Passage

When the sunlight strikes raindrops in the air, they act like a prism and form a rainbow. The rainbow is a division of white light into many beautiful colors. These take the shape of a long round arch, with its path high above, and its two ends apparently beyond the horizon. There is, according to legend, a boiling pot of gold at one end. People look but no one ever finds it. When a man looks for something beyond reach, his friends say he is looking for the pot of gold at the end of the rainbow.

  f. <u>Grimace (G)</u>. The test subject shall grimace, smile, frown, and generally contort the face using the facial muscles. Continue for at least 15 seconds. The test is use to check the reseal of the respirator after seal is broken.

  g. <u>Bend over and touch toes (B)</u>. The test subject shall bend at the waist and touch toes and return to upright position. Repeat for at least one minute.

  h. <u>Jogging in place (J)</u>. The test subject shall perform jog in place for a least one minute.

OSHA Instruction CPL 2-2.54
FEB 10 1992
Office of Science and Technology Assessment

    i.   <u>Normal Breathing (NB)</u>. Same as exercise (a) above.

    j.   The test shall be terminated whenever any single peak penetration exceeds five percent for half-masks and one percent for full facepieces. The test subject may be refitted and retested. If two of the three required tests are terminated, the fit shall be deemed inadequate.

6. **Calculation of Fit Factors**.

    a.   The fit factor determined by the quantitative fit test is expressed as the ratio of challenge concentration outside the respirator to the concentration inside the respirator.

    b.   The average test chamber concentration is the arithmetic average of the test chamber concentration at the beginning and of the end of the test.

    c.   The average peak concentration of the challenge agent inside the respirator shall be the arithmetic average peak concentrations for each of eight exercises of the test which are computed as the arithmetic average of the peak concentrations found for each breath during the exercise.

    d.   The average peak concentration for an exercise may be determined graphically if there is not a great variation in the peak concentrations during a single exercise.

    e.   When fit factors are calculated by a computer the average concentration, instead of average peak concentration, may be used.

B. <u>Qualitative Respirator Fit Test (QLFT)</u>.

    1.   Purpose - The Qualitative Respirator Fit Test procedures in this section shall be performed to supplement quantitative fit testing to meet the semi-annual fit testing requirement for a specific standard such as asbestos or acrylonitrile.

OSHA Instruction CPL 2-2.54
FEB 10 1992
Office of Science and Technology Assessment

2. <u>Respirator Selection</u>.

   a. The test will be performed using the respirator which was determined to be the most effective during the last quantitative fit test. Isoamyl acetate will be the agent used. Individuals must be tested to ensure that they can detect, isoamyl acetate. It may be necessary to use the irritant smoke test, when an individual cannot detect isoamyl acetate.

   b. The fitting process shall be conducted in a room separate from the fit-test room to prevent odor fatigue. Both rooms shall be well ventilated and separated by a distance for enough to avoid cross contamination. Prior to the selection process, the test subject shall be shown how to put on a respirator, how it should be positioned on the face, how to set strap tension, and how to assess a "comfortable" respirator. A mirror shall be available to assist the subject in evaluating the fit and positioning of the respirator. This will not constitute his/her formal training on respirator use, only a review.

   c. Assessment of comfort shall include reviewing the following points with the test subject:

      -Chin properly placed
      -Positioning of mask on nose
      -Strap tension
      -Fit across nose bridge
      -Room for safety glasses
      -Distance from nose to chin
      -Room to talk
      -Tendency to slip
      -Cheeks filled out
      -Self-observation in mirror
      -Adequate time for assessment.

   d. The test subject shall conduct the conventional negative and positive-pressure fit checks (e.g., see American National Standards Practices for Respiratory Protection, ANSI Z88.2-1980). Before

OSHA Instruction CPL 2-2.54
FEB 10 1992
Office of Science and Technology Assessment

conducting the negative or positive-pressure checks, the subject shall be told to "seat" the mask by rapidly moving the head side-to-side and up and down, taking a few deep breaths.

e. The test subject is now ready for fit testing.

f. After passing the fit test, the test subject shall be questioned again regarding the comfort of the respirator. If it has become uncomfortable, another model of respirator shall be tried.

3. **Fit test procedure**.

   a. Each respirator used for fitting and fit testing shall be equipped with organic vapor cartridges or offer protection against organic vapors. The cartridges shall be changed at least weekly.

   b. After selecting, donning, and properly adjusting a respirator, the test subject shall wear it to the fit testing room.

   c. Each test subject shall wear her/his respirator for at least five minutes before starting the fit test.

   d. **Exercise Regimen**.

      (1) **Normal Breathing (NB)**. In the normal standing position, without talking, the subject shall breathe normally for at least one minute.

      (2) **Deep Breathing (DB)**. In the normal standing position the subject shall do deep breathing for at least one minute pausing so as not to hyperventilate.

      (3) **Turning head side to side (SS)**. Standing in place the subject shall slowly turn his/her head from side between the extreme position to each side. Perform for at least one minute.

OSHA Instruction CPL 2-2.54
FEB 10 1992
Office of Science and Technology Assessment

    (4)   <u>Moving head up and down (UD)</u>. Standing in place, the subject shall slowly move his/her head up and down between the extreme position straight up and the extreme position straight down. The head shall be held at each extreme position for at least five seconds. Perform for at least one minute.

    (5)   <u>Reading (R)</u>. The subject shall read out slowly and loud so as to be heard clearly by the test conductor or monitor. The test subject shall read the "rainbow passage."

Rainbow Passage

When the sunlight strikes raindrops in the air, they act like a prism and form a rainbow. The rainbow is a division of white light into many beautiful colors. These take the shape of a long round arch, with its path high above, and its two ends apparently beyond the horizon. There is, according to legend, a boiling pot of gold at one end. People look, but no one ever finds it. When a man looks for something beyond reach, his friends say he is looking for the pot of gold at the end of the rainbow.

    (6)   <u>Bend over and touch toes</u>. The test subject shall bend at the waist and touch toes and return to upright position. Repeat for at least one minute.

    (7)   <u>Jogging in place (J)</u>. The test subject shall perform jog in place for a least one minute.

    (8)   <u>Normal Breathing (NB)</u>. Same as exercise (1) above.

OSHA Instruction CPL 2-2.54
FEB 10 1992
Office of Science and Technology Assessment

    e. If at anytime during the test, the subject detects the odor of the testing agent, she/he shall quickly exit from the test chamber and leave the test area.

    f. If the entire test is completed without the test subject detecting the odor of the testing agent, the test is passed and the respirator selected is judged adequate.

C. **Other Requirements**:

    a. An analysis of the fit test records of our CSHOs in the past shows the median fit factor to be 3,000 for a half-mask. Therefore, the test subject shall not be permitted to wear a half-mask or full facepiece respirator if the minimum fit factor of 500 or 3,000, respectively, cannot be obtained. If hair growth or apparel interferes with a satisfactory fit, then they shall be altered or removed so as to eliminate interference and allow a satisfactory fit. If a satisfactory fit is still not attained, the test subject must use a positive-pressure respirator such as a powered air-purifying respirator, supplied air respirator or self-contained breathing apparatus.

    b. The test shall not be conducted if there is any hair growth between the skin and the facepiece sealing surface.

    c. If a test subject exhibits difficulty in breathing during the tests, she or he shall be referred to a physician trained in respirator disease or pulmonary medicine to determine whether the test subject can wear a respirator while performing her or his duties.

    d. The test subject shall be given the opportunity to wear the assigned respirator for one month. If the respirator does not provide a satisfactory fit during actual use, the test subject may request another quantitative fit test which shall be performed as soon as possible.

OSHA Instruction FEB 1 0 1992
Office of Science and Technology Assessment

    e.   A respirator fit factor card shall be issued to the test subject with the following information:

        (1)   Name.
        (2)   Date of fit test.
        (3)   Make and model of QNFT equipment.
        (4)   Fit factors obtained through each manufacturer, model and approval number (such as TC-21C-XXX) of the respirator tested.
        (5)   Name and signature of the person who conducted the test.

        (6)   Filters used for qualitative or quantitative fit testing shall be replaced whenever increased breathing resistance is encountered, or when the test agent has altered the integrity of the filter media. Organic vapor cartridges/canisters shall be replaced whenever there is any indication of breakthrough by the test agent. Each fresh filter or sorbet cartridge shall be dated when it is installed.

    Note: *The fit test card, OSHA Form 187 is available from the Directorate of Technical Support.*

    f.   1926.58(h)(4)(ii) requires that employees who are wearing negative pressure respirators be fit tested at least every six months.

    g.   In addition, because the sealing of the respirator may be affected, qualitative fit testing shall be repeated immediately and quantitative fit testing as soon as possible when the test subject has:

        (1)   Weight change of 20 pounds or more,
        (2)   Significant facial scarring in the area of the facepiece seal,
        (3)   Significant dental changes, i.e., multiple extractions without prothesis, or acquiring endures,
        (4)   Reconstructive or cosmetic surgery, or
        (5)   Any other condition that may interfere with facepiece sealing.

OSHA Instruction CPL 2-2.54
FEB 10 1992
Office of Science and Technology Assessment

D. **Recordkeeping.**

A summary of all test results shall be maintained in Regional Office for seven years (see OSHA Instruction ADM 12-7.2A). These records shall be considered as **employee exposure records**. A copy of the summary shall include:

1. Name of test subject.

2. Date of testing.

3. Name of the test conductor.

4. Fit factors obtained from every respirator tested (indicate manufacturer, model, size and approval number).

5. Name and type of facepiece(s) which has failed during the qualitative test or has yielded a fit factor less than those prescribed in paragraph F.7.c.(1).

6. When applicable, the strip chart for fit testing shall also be maintained for seven years (see OSHA Instruction ADM 12-7.2A), and should contain the following information:

    a. Date and time
    b. Name of test subject
    c. Name of test conductor
    d. Manufacturer name and model of QNFT equipment
    e. Respirator (type, brand, style, size and TC number)
    f. Chart scale (e.g. 1 percent-full scale)
    g. Chart speed
    h. Initial ambient
    i. Initial base line estimate
    j. Individual exercise (mark beginning and end)
    k. Final base line
    l. Calculation of penetration for each exercise
    m. Overall fit factor (average of all exercises)
    n. Notes (eyeglasses, dentures, scars, etc.)

Copies of all records shall be forwarded to Directorate of Technical Support annually in the form of computer diskette(s), with the Regional Program Evaluation report.

OSHA Instruction CPL 2-2.54
FEB 10 1992
Office of Science and Technology Assessment

Chapter VI

Maintenance

A. A program for respirator cleaning and care shall be established as a part of the standard operating procedures (SOP) by all Regional and Area Offices. The purpose of this element of a respirator program is to assure that all respirators are properly maintained. If they are modified in any way, their protection may be reduced. The Program Coordinators shall be trained to inspect, clean, repair, and store respirators. The program should be based on the number and types of respirators, working conditions, and hazards involved.

In general, the program should include: inspection, cleaning, repair and storage.

1. All respirators shall be inspected before and after each use. The cleaning of respirators is the responsibility of the CSHO who is using the respirator, not the Program Coordinator.

2. SCBAs and all Emergency Egress Respirators shall be inspected on a monthly basis and before or after each use to assure they will perform satisfactorily. Inspection records shall be maintained for each SCBA/Emergency Egress for seven years (see OSHA Instruction ADM 12-7.2A). At a minimum, these records will include: date, inspector, and any unusual conditions or findings. Any repairs or modification to these units shall be documented in detail.

3. Air-purifying respirators:

   a. Thoroughly check all connections, gaskets and valves for proper fit and tightness. Check the condition of the facepiece and all its parts, and all connecting air tubes and head bands. Inspect parts, and all connecting air tubes and head bands. Inspect rubber or elastomer parts for pliability and signs or deterioration.

   b. Clean and disinfect respirators as follows:

      (1) Remove all cartridges, canisters, and filters, plus gaskets or seals not affixed to their seats. Cartridges will be discarded.

OSHA Instruction CPL 2-2.54
FEB 10 1992
Office of Science and Technology Assessment

       (2) Remove elastic head bands.

       (3) Remove exhalation cover.

       (4) Remove speaking diaphragm or speaking diaphragm-exhalation valve assembly.

       (5) Remove inhalation valves.

       (6) Wash facepiece and breathing tube in cleaner/sanitizer recommended by the manufacturer with warm water, use manufacturers' recommended temperature. Wash components separately from the facepiece, as necessary. Remove heavy soil from surfaces with a hand brush.

   c. After respirators have been inspected, cleaned, sanitized, and repaired, store them so as to protect against dust, excessive moisture, damaging chemicals, extreme temperatures and direct sunlight.

   d. Each unit shall be sealed in a plastic bag, placed in a separate box, and tagged for immediate use.

   e. Cartridges and canisters shall always be stored in their sealed plastic bags until ready for use. Canisters will be stored with original seals intact in the upright position.

4. Examples of inspection procedures for the commonly used Mine Safety Appliance (MSA) and the Survivair SCBAs are listed in Appendices A and B.

OSHA Instruction CPL 2-2.54
FEB 10 1992
Office of Science and Technology Assessment

APPENDIX A

SAMPLE CHECKLIST AND INSPECTION FORM

FOR MSA SELF-CONTAINED BREATHING APPARATUS

Video tapes reviewing proper inspection and maintenance procedures for MSA and other SCBAs are available at the Regional office.

I. <u>Backpack and Harness Assembly</u>

    A. **Visually inspect**

        1. Straps and buckles -- in place and not worn.

        2. Back plate and cylinder lock -- not cracked or broken.

II. <u>Cylinder and Valve Assembly</u>

    A. **Pressure gauge**

        1. Leaks.

        2. Broken lens or needle.

        3. If gauge reads less than 90 percent of rated capacity have the cylinder refilled.

    B. **Cylinder valve**

        1. If present, make sure lock is functional. Cylinder lock is no longer required, and may be absent on some units.

III. <u>Regulator and Hose Assembly</u>

    A. Hose and Connector

        1. <u>Connector</u>

            a. Inspect "O-Ring" seal (in place and in good condition).

            b. Connection tight on cylinder.

OSHA Instruction CPL 2-2.54
FEB 10 1992
Office of Science and Technology Assessment

    2. <u>Hose</u>

        a. No leaks (cylinder valve open).

        b. No bubbles in outer coating (cylinder valve open).

B. **Regulator**

    1. Mainline valve (gold color) <u>closed</u>.

    2. Bypass valve (red color) <u>closed</u>.

    3. Open cylinder valve (Audi-Larm should ring).

    4. Cover regulator outlet with palm of hand and open mainline valve (gold).

    5. Check pressure reading on regulator pressure gauge.

    NOTE: Cylinder should be full (2216 psig) or contain a sufficient quantity of air to perform the anticipated task. Cylinders containing 90 percent of rated capacity or less air pressure <u>MUST BE REFILLED</u>.

    6. In rapid succession, remove and replace the palm of hand over the regulator outlet. Air flow should be present when hand is removed from outlet, and should be completely stopped when outlet is covered.

    7. Keep regulator outlet covered and close cylinder valve.

    8. Slowly remove hand from regulator outlet and observe the needle on the regulator pressure gauge. Audi-Larm should ring between 500 and 600 psig.

    9. Check the integrity of the regulator diaphragm by exhaling into the regulator outlet. If the diaphragm is seated properly and has no defects, it <u>will not</u> be possible to force the exhaled air through the regulator. In addition, a negative pressure test on the regulator should be performed.

OSHA Instruction CPL 2-2.54
FEB 10 1992
Office of Science and Technology Assessment

10. Close main-line (gold) valve

11. Open cylinder valve

12. Making sure there are no obstructions at the regulator outlet, gently open the bypass (red) valve. Air should easily flow from the regulator outlet, thus assuring that the bypass valve is functional

    NOTE: During this portion of the checkout procedure, there will be no indication of pressure on the regulator pressure gauge. This condition is due to the fact that the mainline valve is closed, and no pack pressure on the regulator outlet may be applied when the bypass valve is opened

13. Allow all residual air to escape through the bypass valve, then close the bypass (red) valve. Audi-Larm should ring at 500 to 600 psig, but there will be no indication of pressure on the gauge as previously discussed

IV. Facepiece Assembly

   A. **Breathing tube and connector**

      1. Breathing tube

         a. Stretch and visually inspect for holes or deterioration

      2. **Connector**

         a. Check threads for damage

         b. Check rubber gasket (in place and in good condition)

   B. **Facepiece**

      1. Head Harness

         a. Inspect end stops and serration for damage and deterioration

OSHA Instruction CPL 2-2.54
FEB 10 1992
Office of Science and Technology Assessment

       b. Check all rubber for deterioration

  2. **Lens**

       a. Inspect for scratches, cracks and proper mounting in the facepiece.

  3. **Exhalation Valve**

       a. Locate exhalation valve stem, inside respirator at the bottom. With index finger, press down on valve stem. This action will free the valve if it has become stuck in the closed position during storage.

  4. **Overall Facepiece**

       a. Inspect for deterioration or holes.

<u>General</u>

A. Ascertain all straps are straight and functional, and buckles and snaps are in operating condition. Then don backpack.

<u>Sign-off</u>

A. Date, sign, mark psig, and findings on inspection sticker.

B. Forward one (1) copy of completed form to AA-TS.

      (Source: Los Alamos National Laboratory)

OSHA Instruction CPL 2-2.54
FEB 1 0 1992
Office of Science and Technology Assessment

SAMPLE
SCBA INSPECTION FORM   AREA OFFICE_____

                              DATE_____

                              SCBA IDENTIFICATION_____

Place a check mark in appropriate space and note any defects:

Inspection Point: *Satisfactory*   *Unsatisfactory*   *Comments*

1. Straps and Buckles
2. Back Plate and Cylinder Lock
3. Pressure Gauge (Cylinder)
4. Cylinder Pressure
5. Cylinder Valve Lock
6. High Pressure Hose
7. High Pressure Hose Connector
8. By-Pass Valve
9. Main-Line Valve
10. Pressure Gauge (Regulator)
11. Pak-Alarm
12. Breathing Tube
13. Breathing Tube Connector
14. Breathing Tube Gasket
15. Facepiece Body
16. Head Harness
17. Lens
18. Exhalation Valve

                                              Inspected by:_____

Forward one (1) copy of completed form
to AA-TS. (Source: Los Alamos National Laboratory)

OSHA Instruction CPL 2-2.54
FEB 10 1992
Office of Science and Technology Assessment

APPENDIX B

SAMPLE INSPECTION AND MAINTENANCE PROCEDURES FOR SURVIVAIR SELF-CONTAINED BREATHING APPARATUS

The SCBA should be inspected for defects before and after each use, and at least once monthly if it not used. Units that are used for Emergency Response should be inspected and functional tested daily. The SCBA should be repaired as necessary, cleaned and disinfected, and then stored properly to assure that is it maintained in satisfactory working condition. A record should be kept of inspection dates and findings.

VISUAL INSPECTION

The following is a recommended checklist for visually inspecting the Mark 2 SCBA.

I. **Facepiece**

   A. Facepiece Lens - Watch For:

      1. Nicks, scratches, or abrasion which would impair outward visibility.
      2. Deep gouges or cracks which would reduce impact resistance.

   B. Headstrap Buckles

      1. Crushed, bent, or corroded buckles.
      2. Damaged or loosened rivets securing buckles to facepiece.

   C. Mask Rims

      1. Tightness of the rim screws.
      2. Deformation or broken rim.

   D. Inlet Nozzle

      1. Screws securing the nozzle cover.
      2. Clamp securing the low pressure hose to the nozzle.

OSHA Instruction CPL 2-2.54
FEB 10 1992
Office of Science and Technology Assessment

    3. Exhalation valve seat (in nozzle).
    4. Heat damage to the nozzle body (warping, cracking, etc.)
    5. Nozzle spring.
    6. Exhale through the exhalation valve, no sticking should be evident.

  E. <u>Facepiece Skirt</u>

    1. Cuts, gouges, or punctures.
    2. Tears or nicks in sealing lips.
    3. Age or heat induced damage to the rubber.
    4. Pull out of attaching rivets.

  F. <u>Convoluted Low Pressure Hose</u>

    1. Cuts, nicks or punctures in the hose.
    2. Age or heat induced cracking, crazing, or hardening of the rubber.
    3. Crushed, broken, or cracked quick connect.

  G. <u>Headstrap Spider</u>

    1. Abrasion or nicking of headstrap legs.
    2. Age or heat induced hardening of the rubber.

II. Second Stage Regulator

    1. Case and Cover - Dents or heat damage.
    2. Silicone "Trap Door" valve assembly.
    3. "O" ring in low pressure air outlet.
    4. Smooth operation of bypass valve.
    5. Damaged threads and worn out slots on "quick on" adaptor.
    6. Pressure gauge - lens clarity, pointer to zero, deformation of needle.
    7. Pressure gauge hose and fittings.

III. Audio Alarm, First Stage Regulator, and Intermediate Pressure Hose

    1. Debris or water under the bell.
    2. Tightness of the socket head screws securing bell to body.
    3. Dents or deformation of bell.
    4. Hose and end connectors.
    5. Retaining rings secure on end connectors.

OSHA Instruction CPL 2-2.54
FEB 10 1992
Office of Science and Technology Assessment

      6. Abrasion of hose.
      7. Condition of female threads on C.G.A. handwheel.
      8. Condition of o-ring and groove on C.G.A. nipple.

IV. **Backpack and Air Cylinder**

    A. <u>Backpack</u>

        1. Proper function of cylinder - securing band and latch.
        2. Cylinder secure in frame and band.
        3. Frame bent or broken.
        4. Stitching of webbing.
        5. Buckles - shoulder straps and waist belt.

    B. <u>Air Cylinder</u>

        1. External inspection - dents, gouges, blisters, or discolored paint.
        2. Internal inspection - every three years inspect interior of cylinder.
        3. Last hydrostatic test date - every three years.
        4. External damage to cylinder valve.
        5. Smooth operation of valve handwheel and ratchet mechanism.
        6. Tightness of screws securing rubber guard on cylinder valve.
        7. Condition of threads on valve outlet.
        8. Cylinder pressure - gauge needle in the green area.

OSHA Instruction CPL 2-2.54
FEB 10 1992
Office of Science and Technology Assessment

NOTE

If any defects are found or if anything appears to be different or unusual from the norm, <u>do not</u> use the equipment until it has been checked by an authorized repair technician.

FUNCTIONAL TESTING
PERFORMANCE AFTER CLEANING AND PRIOR TO RETURNING TO SERVICE

I. Facepiece

  1. Don the facepiece.
  2. Block inlet of low pressure hose with palm of hand.
  3. Inhale gently - facepiece should "collapse."
  4. Exhale through the exhalation valve, no sticking should be evident.

II. Regulator

  1. Attach regulator to fully charged cylinder.
  2. Close bypass valve and slowly open cylinder valve.
  3. Check that cylinder valve and regulator pressure gauge show the same.
  4. Attach facepiece and inhale - regulator should deliver on inhalation without excessive effort, free flow, or fluttering.
  5. Slowly open bypass valve. A steady flow of air should enter the facepiece. No delay in air flow.
  6. Turn "Quick-On" fitting all the way to the left - no air should flow when the facepiece is off.

III. Leak Test

  1. Open cylinder valve to fully pressurize the regulator - close cylinder valve.
  2. Wait 15 seconds. While looking at the regulator gauge, open the cylinder valve. Any movement indicates a leak.

OSHA Instruction CPL 2-2.54
FEB 10 1992
Office of Science and Technology Assessment

IV. <u>Audio Alarm Test</u>

1. Remove the facepiece from the regulator.
2. Open cylinder valve to fully pressurize the regulator - close cylinder valve.
3. Slowly open bypass valve and watch the regulator pressure gauge. Alarm should begin ringing when the gauge reads 1/4 full.
4. Bleed all residual air from the system using bypass valve, the close valve completely. Complete inspection record.

CLEANING

1. For sanitary reasons, the facepiece should be cleaned and disinfected after each use, even if it does not look dirty.

2. Make a cleaning solution by mixing water with any detergent that contains effective disinfectants (such as quaternary ammonium compounds.)

3. Heat the solution to 140-160 F.

4. Immerse the facepiece in the cleaning solution.

5. Using a soft brush, gently clean the facepiece.

6. Rinse the facepiece in a fresh-water bath and allow it to air dry. Mild heat (less than 160 F) may be used to speed up the drying. Use of a towel to dry the facepiece is not recommended unless a clean, lint-free towel is used.

7. Use caution when cleaning the facepiece lens. Although the outer surface of the lens has a proprietary anti-scratch coating, it can be damaged through careless or abusive handling. do not attempt to "polish out" scratches with any abrasive agent as this will only cause further damage to the remaining coating. Warm, soapy water (using Joy, Mr. Clean, Simple Green, Lestoil, etc.) is usually adequate to remove adhering grime. Stubborn deposits may require the use of denatured or isopropyl alcohol or other mild solvents. Do not allow any solvent to come into contact with rubber or plastic parts. Use solvents only in a well-ventilated area.

8. Apply Survivair anti-fog PN 9510-15 to the inside of the lens.

INDEX

Acid Gases 5, 18, 26
Acid Gases and Ammonia Gas 18, 26
Acid Gases and Organic Vapors 18, 26
Acid Gases, Organic Vapors
    and Ammonia Gases 18, 26
Air Quality 22
Ammonia Gas 18, 26
Atmosphere Contaminants 18-19, 25-26
Colors Assigned 18-19, 26-27
Carbon Monoxide 18, 26
Chlorine Gas 18, 26
Copy of Respiratory Protection Standard .. 22-25
Compliance Safety and Health Officers
    (CSHO) 37-38, 41-42, 46
Emergency Escape Respirators 48
Emergency Use Respirators
Employee Training 13-14
    Discipline1 3-14
    Exercise Regime 56-57, 60-61
    Instruction1 3-14
Fit Testing Respirators 7, 14, 23-24, 52, 60
    General Requirements 53-54
    Half-Mask Respirators 54
    Isoamyl Acetate Test 55, 59
    Irritant Fume Test 55-56
    Negative Pressure Test 55, 59, 63
    Positive Pressure Test 55, 59
    Procedural Reqs 54
    Quantitative Respirator Fit Test
        (QNFT) 7-8, 20, 52, 56
    Qualitative Respirator Fit Test
        (QLFT) 7, 20, 52, 58, 63
    Test Chamber 53-54
Gas and Vapor Contaminants 2
Gas Mask Table 18-19
Gaseous and Particulate Contaminants 12
Gas Mask Cannister 17-18, 24-27
Hydrocyanic Acid Gas 18, 26
Hydrocyanic Acid Gas and Chloropicrin Vapors
    .... 18, 26
Instructions For Use in the Respiratory Protection
    Program .... v-vii, 14
Introduction to Respiratory Protection Program . 2

Mine Safety and Health Administration
    (MSHA) 9, 22
National Institute for Occupational Safety
    (NIOSH) 9, 22
Organic Vapors 18, 26
Oxygen Deficiency 12
OSHA Instruction CPL 31, 54
    Respiratory Protection Program Manual . 131
Particulates (Dusts, Fumes, Mists, Fogs, or
Smokes) in combination with any of the
    above gases or vapors 19, 26
Particulate Contaminants 12
Radioactive Materials, Excepting Tritium and
    Noble Gases 19, 26
Respiratory Protection Program Administration ...
    .... 2-3, 39
    Cleaning 15-16, 22, 40, 65, 67
    Emergency Use 16-18
    Inspection 15-16, 22, 24, 40, 65, 67
    Maintenance 65
    Medical Questionnaire .... v, 5, 8
    Recordkeeping 64
    Repair 15-16
    Storage 15-16, 24
    Training 51
    Use of Respirators 3-9, 23, 43
Respirator Protection Standard 22-25
Respirator User Card .... v, 6
Selection of Respirators
    .... 9-11, 22, 39-40, 43, 49, 59
    Air Purifying Respirators
    .... 4, 10-11, 44, 46, 48, 53, 65
    Atmosphere Supplying Respirators ... 10, 46
    Inspection Form 71
    Inspection and Maintenance Procedures 72-76
    Self-Contained Breathing Apparatus
        (SCBA) 16, 23-24, 39, 46-50, 65-70
Training Record For Respiratory Protection
Program .... v, 20
    Use of Respirators 3-9, 23
    U.S. Dept. of Health and Human Services . 9
    U.S. Dept. of Labor 9

About Government Institutes

Government Institutes, Inc. was founded in 1973 to provide continuing education and practical information for your professional development. Specializing in environmental, health and safety concerns, we recognize that you face unique challenges presented by the ever-increasing number of new laws and regulations and the rapid evolution of new technologies, methods and markets.

Our information and continuing education efforts include a Videotape Distribution Service, over 200 courses held nation-wide throughout the year, and over 250 publications, making us the world's largest publisher in these areas.

Government Institutes, Inc.
4 Research Place, Suite 200
Rockville, MD 20850
(301) 921-2355

Other related books published by Government Institutes:

Health Effects of Toxic Substances   This comprehensive book provides you with an excellent understanding of the toxicology and industrial hygiene of hazardous materials. Chapters cover: Industrial Toxicology - History and Hazards; Exposure and Entry Routes - Pharmacokinetics I; Distribution, Localization, Biotransformation, Elimination; Dose-Effects and Time-Effects Relationships; Classification, Type, and Limits of Exposure; Action of Toxic Substances Pharmacodynamics; Target Organ Effects; Reproductive Toxins, Mutagens, and Carcinogens; Survey of Common Hazardous Agents I, Toxic Substances; Survey of Common Hazardous Agents II, Physical & Biological Hazards; Types of Environmental Health Hazards; Monitoring of Harmful Agents; Exposure Limits and Personal Protective Equipment; Exposure Control Methods; Medical Monitoring, Treatment, and Management; Risk Assessment. *Softcover, Index, 300 pages, Aug. '95, ISBN: 0-86587-471-9 $39*

Understanding Workers' Compensation: *A Guide for Safety and Health Professionals*   This book explains in simple and direct terms the Workers' Compensation System. It provides a basic understanding of injury prevention, types of injuries, and cost containment strategies. This book includes sample forms, checklists for work site evaluations, and an appendix containing material from the most recent U.S. Chamber of Commerce analysis, comparisons of all state and Canadian provincial laws, policies on rehabilitation, statistics on benefits payable, and waiting periods. A directory of state and provincial workers compensation administrators with full contact information is also included. *Softcover, 192 pages, June '95, ISBN: 0-86587-464-6 $45*

So You're the Safety Director: An Introduction to Loss Control and Safety Management   Author Michael Manning, a safety veteran and sought-after consultant and speaker, has created an introduction to your bottom-line responsibilities, concentrating on your role in evaluating, managing, and controlling your company's losses and handling the OSHA compliance process. Manning's narrative approach and easy-to-follow writing style make it seem like you've hired him to help you start — or upgrade — your safety program, which is exactly what hundreds of companies have done. Let Manning walk you through the in's and out's of establishing and evaluating your company's safety program: comparing your safety program to those of similar companies, establishing safety committees, involving all employees in your safety program, investigating accidents and preventing their recurrence, managing your compensation costs, preparing for and handling OSHA inspections, and using your company's insurance company as a resource. *Softcover/Index/184 pages/Oct '95/$45   ISBN: 0-86587-481-6*

Ergonomic Problems in the Workplace:A Guide to Effective Management   The valuable insights you'll gain from this new book will help you develop and implement your own successful ergonomics program. Now your company can reduce injuries — such as Cumulative Trauma Disorders (CTD) — and reduce the number of workers' compensation claims. In addition, case studies help you learn from the successes and failures of other companies. Table of contents includes: developing an ergonomics program; management commitment; case histories; hazard assessment; cumulative trauma disorders; workplace hazards; hazard prevention and controls; back injuries and material handling; tool selection; ergonomic personal protective equipment; implementing an ergonomics program; medical management; VDTs and office ergonomics; heat stress; training; ADA and ergonomics; working with OSHA on ergonomic issues; and sources of information and assistance. *Softcover/272 pages/Sept '95/$59 ISBN: 0-86587-474-3*

Call the above number for our current book/video catalog and course schedule.

Publications (cont.)

OSHA Field Inspection Reference Manual — This new revision of inspection guidelines, previously contained in the OSHA Field Operations Manual, is now being used by OSHA inspectors when checking your facility for compliance. Learn where the inspectors will look, what they'll look for, how they'll evaluate your working conditions, and how they'll actually proceed once inside your facility. *Softcover, 144 pages, Jan '95, ISBN: 0-86587-426-3* **$59**

OSHA Technical Manual, 3rd Edition — This OSHA inspection manual includes chapters on: Personal Sampling Techniques and Procedures for Air Contaminants; Sampling for Surface Contamination; Heat Stress; Noise Measurement; Back Disorders and Injuries in Industry; Indoor Air Quality Investigations; Hospital Investigations: Health Hazards; Technical Equipment for Testing and Monitoring; Shipping and Handling of Samples; Pressure Vessel Guidelines; Demolition; Chemical Protective Clothing; Oilwell Derrick Stability; Industrial Robots and Robot System Safety; and more. *Softcover, 300 pages, Nov '93, ISBN: 0-86587-366-6* **$79**

Safety Made Easy: A Checklist Approach to OSHA Compliance   Written by Tex Davis, this book provides a new, simpler way of understanding your requirements under the complex maze of OSHA's workplace safety and health regulations. The easy-to-use format and logical organization make this book ideal for those who are just entering the field of safety compliance as well as for experienced safety professionals.   *Softcover, 180 pages, May '95, ISBN: 0-86587-463-8*   **$45**

Written Compliance Programs are the cornerstone of compliance with OSHA standards, and are always requested by OSHA inspectors. Creating them from scratch is a laborious task, but by using the Wordperfect diskette in this book-disk package, users can customize these boilerplate programs and produce their own company-specific written program, quickly and easily.

Electrical Safety and Lockout/Tagout: *Proven Written Programs for Compliance*
Details what types of work are covered and excluded, compliance procedures, and training.   *Softcover w/ WordPerfect disk, approx. 170 pages, Nov '95, ISBN: 0-86587-502-2* **$59**

OSHA Hazard Communication Standard: A Proven Written Program for Compliance
This package will ensure you have a written program in place to comply with the single-most frequently cited OSHA violation.   *Softcover w/Wordperfect disk, approx. 160 pages, Nov '95, ISBN: 0-86587-499-9* **$59**

OSHA's Process Safety Management Standard: *A Proven Written Program for Compliance*
Features 12 easy-to-use checklists for performing a self-audit of your compliance status, and a convenient, reader-friendly list of the Toxic and Reactive Chemicals regulated by the standard together with the amount of each that must be on hand in order to trigger the standard's coverage. *Softcover w/ WordPerfect disk, approx. 150 pages, Nov '95, ISBN: 0-86587-500-6* **$59**

Educational Programs

■ Our **COURSES** combine the legal, regulatory, technical, and management aspects of today's key environmental, safety and health issues — such as environmental laws and regulations, environmental management, pollution prevention, OSHA and many other topics. We bring together the leading authorities from industry, business and government to shed light on the problems and challenges you face each day. Please call our Education Department at (301) 921-2345 for more information!

■ Our **TRAINING CONSULTING GROUP** can help audit your ES&H training, develop an ES&H training plan, and customize on-site training courses. Our proven and successful ES&H training courses are customized to fit your organizational and industry needs. Your employees learn key environmental concepts and strategies at a convenient location for 30% of the cost to send them to non-customized, off-site courses. Please call our Training Consulting Group at (301) 921-2366 for more information!

License Agreement

By opening this package you indicate your acceptance of the **Government Institutes, Inc.** software licensing agreement with the following terms:

Single Users: For each licensed use of the Software which you have purchased, only one person may access the Software at any given time.

Restrictions: You may not and you may not permit others to (a) disassemble, decompile or otherwise derive source code from the Software, (b) reverse engineer the Software, (c) modify or prepare derivative works of the Software, (d) copy the Software, except to make a single copy for archival purposes only, (e) rent or lease the Software, (f) use the Software in an online system, (g) use the Software in any manner that infringes the intellectual property or other rights of another party, or (g) transfer the Software or any copy thereof to another party, unless you transfer all media and written materials in this package and retain no copies of the Software (including prior versions of the Software) for your own use.

Limited Warranty and Limitation of Liability: For a period of 60 days from the date the Software is acquired by you, Government Institutes warrants that the media upon which the Software resides will be free of defects that prevent you from loading the Software on your computer. Government Institutes' sole obligation under this warranty is to replace any defective media, provided that you have given Government Institutes notice of the defect within such 60-day period. The Software and data are licensed to you on an **"AS IS"** basis without any warranty of any nature.

Government Institutes disclaims all other warranties, express or implied, including the implied warranties of merchantability and fitness for a particular purpose. Government Institutes shall not be liable for any damage or loss of any kind arising out of or resulting from your possession or use of the software and data (including data loss or corruption). In addition, the authors, editors, and publisher assume no liability of any kind whatsoever resulting from the use of or reliance upon the contents of this product. Regardless of whether such liability is based in tort, contract or otherwise. If the foregoing limitation is held to be unenforceable, Government Institutes' maximum liability to you shall not exceed the amount of the license fees paid by you for the software. The remedies available to you against Government Institutes under this agreement are exclusive. Some states do not allow the limitation or exclusion of implied warranties or liability for incidental or consequential damages, so the above limitations or exclusions may not apply to you.